Audio IC Circuits Manual

Audio IC Circuits Manual

R. M. Marston

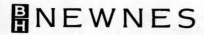

BH NEWNES

To Esther, with love

Butterworth-Heinemann Ltd
Linacre House, Jordan Hill, Oxford OX2 8DP

℞ A member of the Reed Elsevier group

OXFORD   LONDON   BOSTON
MUNICH   NEW DELHI   SINGAPORE   SYDNEY
TOKYO   TORONTO   WELLINGTON

First published 1989
Reprinted 1990, 1993

British Library Cataloguing in Publication Data
Marston, R. M. (Raymond Michael), *1937–*
    Audio IC circuits manual
    1. Electronic equipment. Audio integrated circuits
    I. Title
    621.381'73

ISBN 0 7506 1922 8

Printed and bound in Great Britain by
Biddles Ltd, Guildford and King's Lynn

Contents

Preface

A vast range of audio and audio-associated integrated circuits (ICs) are readily available for use by amateur and professional design engineers and technicians. This book is a single-volume guide to the most popular and useful of these devices. It deals with ICs such as low frequency linear amplifiers, dual pre-amplifiers, audio power amplifiers, charged-coupled device (CCD) delay lines, bar-graph display drivers, and power supply regulators, and shows how to use these devices in circuits ranging from simple signal conditioners and filters to complex graphic equalizers, stereo amplifier systems, and echo/reverb delay line systems, etc.

The book is specifically aimed at the practical design engineer, technician, and experimenter, as well as the electronics student and amateur, and deals with the subject in an easy-to-read, down-to-earth, non-mathematical but very comprehensive manner. Each chapter deals with a specific class of IC, and starts off by explaining the basic principles of its subject and then goes on to present the reader with in-depth looks at practical ICs and their applications.

The book is split into seven distinct chapters. Chapters 1 to 4 deal with pure 'audio' subjects, such as audio processing circuits, audio pre-amplifier circuits, and audio power amplifier circuits. Chapters 5 and 6 deal with the audio-associated subjects of light-emitting diode (LED) bar-graph displays (which can give a visual indication of signal levels, etc.) and CCD delay-line circuits (which can be used to give special sound effects such as echo and reverberation, etc.). Chapter 7 deals with power supply circuits for use in audio systems.

Throughout the book, great emphasis is placed on practical 'user' information and circuitry, and the book abounds with useful circuits and graphs; a total of over 240 diagrams are included. Most of the ICs and other devices used in the practical circuits are modestly priced and readily available types, with universally recognized type numbers.

1 Audio processing circuits

An audio processor can be defined as any circuit that takes an audio input signal and generates an audio output that is directly related to that input. The processor may, for example, simply invert the original signal and/or provide it with a fixed amount of voltage gain, as in the case of a linear amplifier; or it may give it an amount of gain that varies with signal frequency, as in the case of an active filter; or an amount of gain that varies with the signal's mean amplitude, as in the case of a 'constant-volume' or non-linear amplifier or an expander/compressor circuit, etc. Practical examples of all these types of processor circuit are presented in this chapter.

The first part of this chapter deals with general-purpose audio-frequency processor circuits based on standard operational amplifier (op-amp) ICs. Next, it deals with operational transconductance amplifier (OTA) ICs, which can be used in a variety of voltage controlled amplifier (VCA) circuits, etc. It ends by looking at two dedicated VCA ICs and their application circuits. Dedicated audio pre-amplifier or power amplifier ICs are not described since they are dealt with in great detail in later chapters of this book.

Op-amp basics

The best known and most versatile type of audio signal processing IC is the ordinary operational amplifier, which can be simply described as a high-gain direct-coupled voltage amplifier block with a single output terminal but with both inverting and non-inverting high impedance input terminals, thus enabling the device to function as either an inverting, non-inverting, or differential amplifier. *Figure 1.1(a)* shows the circuit symbol of the conventional op-amp.

Op-amps are very versatile devices. When coupled to suitable feedback networks they can be used to make precision ac and dc amplifiers, active filters, oscillators, and voltage comparators, etc. They are normally powered from split supplies (as shown in *Figure 1.1(b)* with positive (+ ve), negative (− ve) and common (zero volt) supply rails, enabling the op-amp output to swing either side of the zero volts value and to be set at zero volts when the differential input voltage is zero. They can, however, also be powered from single-ended supplies, if required.

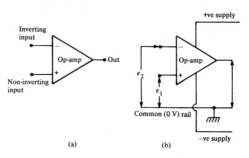

Figure 1.1   *(a) Symbol and (b) supply connections of a conventional op-amp*

Note that the output signal voltage of an op-amp is proportional to the *differential* signal voltage between its two input terminals and, at low audio frequencies, is given by:

$$e_{out} = A_0(e_1 - e_2),$$

where A_0 is the op-amp's low frequency open-loop voltage gain (typically 100 dB, or × 100,000), e_1 is the non-inverting terminal input signal voltage, and e_2 is the inverting terminal input signal voltage.

Thus, an op-amp can be used as a high-gain inverting ac amplifier by grounding its non-inverting terminal and feeding the input signal to its inverting pin via C_1 and R_1, as in *Figure 1.2(a)*, or as a non-inverting ac amplifier by reversing the two input connections as in *Figure 1.2(b)*, or as a differential amplifier by feeding the two input signals to the op-amp as shown in *Figure 1.2(c)*. Note in the latter case that if both input signals are identical the op-amp should, ideally, give zero output signal.

The voltage gains of the *Figure 1.2* circuits depend on the open-loop gains of individual op-amps and on input signal frequencies. *Figure 1.3*, for example, shows the typical frequency response graph of the well known 'type 741' op-amp; its voltage gain is greater than 100 dB at frequencies below 10 Hz, but falls off at a 6 dB/octave (20 dB/decade) rate at frequencies above 10 Hz, reaching unity (0 dB) at an f_T 'unity gain transition' frequency of 1 MHz.

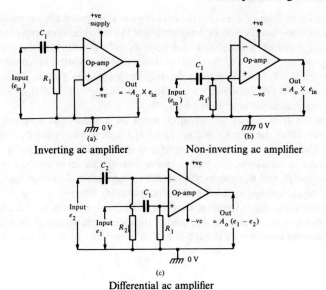

(a)
Inverting ac amplifier

(b)
Non-inverting ac amplifier

(c)
Differential ac amplifier

Figure 1.2 *Methods of using the op-amp as a high-gain open-loop ac amplifier*

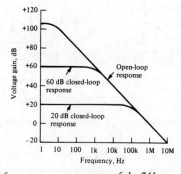

Figure 1.3 *Typical frequency response curve of the 741 op-amp*

This graph is typical of most op-amps, although individual types may offer different A_0 and f_T values.

Closed-loop amplifiers

The best way of using an op-amp as an ac amplifier is to wire it in the closed-loop mode, with negative feedback applied from output to input as shown in

4 Audio processing circuits

the circuits of *Figure 1.4*, so that the overall gain is determined by the external feedback components' values, irrespective of the individual op-amp characteristics (provided that the open-loop gain, A_0, is large relative to the closed-loop gain, A). Note from *Figure 1.3* that the signal bandwidth of such circuits equals the IC's f_T value divided by the circuit's closed-loop A value. Thus, the 741 gives a 100 kHz bandwidth when the gain is set at $\times$ 10 (20 dB), or 1 kHz when the gain is set at $\times$ 1000 ($= 60$ dB).

Figure 1.4(a) shows the op-amp wired as a fixed-gain inverting ac amplifier. Here, the circuit's voltage gain (A) is determined by the R_1 and R_2 ratios and equals R_2/R_1, and its input impedance equals the R_1 value; the circuit can thus easily be designed to give any desired values of gain and input impedance, R_1 and R_2 have no effect on the voltage gain of the actual op-amp, so the signal voltage appearing at its output is A_0 times greater than that appearing on its input terminal; consequently, the signal current induced in R_2 is A_0 times greater than that caused by the input terminal signal alone, and this terminal thus acts as though it has an impedance of R_2/A_0 connected

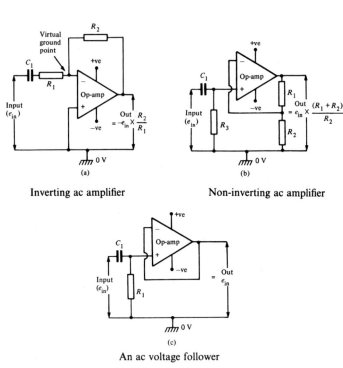

Figure 1.4   *Closed-loop ac amplifier circuits*

between the terminal and ground; the terminal thus acts like a low impedance 'virtual ground' point.

Figure 1.4(b) shows how to connect the op-amp as a fixed-gain non-inverting ac amplifier. In this case the voltage gain equals $(R_1 + R_2)/R_2$. The input impedance, looking into the op-amps's input terminal, equals $(A_0/A)Z_{in}$ where Z_{in} is the open-loop input impedance of the op-amp; this impedance is shunted by R_3, however, so the input impedance of the actual circuit is less than the R_3 value.

The above circuit can be made to function as a precision ac voltage follower by wiring it as a unity-gain non-inverting amplifier, as shown in *Figure 1.4(c)*, where the op-amp operates with 100% negative feedback. The op-amp input impedance is very high in this circuit (roughly $A_0 \times Z_{in}$), but is shunted by R_1, which thus determined the circuit's input impedance value.

Practical op-amps

Practical op-amps are available in a variety of types of IC construction (bipolar, MOSFET, JFET, etc.), and in a variety of types of packaging styles (plastic DIL (dual in-line), metal-can TO5, etc.). Some of these packages house two or four op-amps, all sharing common supply line connections. *Figure 1.5* gives the parameter and outline details of eight popular 'single' op-amp types, all of which use 8-pin DIL packaging.

The 741 and NE531 are bipolar types. The 741 is a very popular general-purpose type featuring internal frequency compensation and full overload protection on inputs and output. The NE531 is a high-performance type with a very high 'output slew rate' capability; an external compensation capacitor (100 pF), wired between pins 6 and 8, is needed for stability, but can be reduced to a very low value (1.8 pF) to give a very wide bandwidth at high gain.

The CA3130 and CA3140 are MOSFET-input op-amps that can operate from single or dual power supplies, can sense inputs down to the negative supply rail value, have very high input impedances (1.5 million MΩ), and have outputs that can be strobed. The CA3130 has a CMOS output stage; an external compensation capacitor (typically 47 pF) between pins 1 and 8 permits adjustment of bandwidth characteristics. The CA3140 has a bipolar output stage and is internally compensated.

The LF351, 411, 441 and 13741 are JFET type op-amps with very high input impedances. The LF351 and 411 are high performance types, while the LF441 and 13741 are general-purpose types that can be used as direct replacements for the popular 741. Note that the LF441 quiescent current consumption is less than one tenth of that of the 741.

Parameter	Bipolar op-amps		MOSFET op-amps		JFET op-amps			
	741	NE531	CA3130E	CA3140E	LF351	LF411	LF441	LF13741
Supply voltage range	±3 V to ±18 V	±5 V to ±22 V	±2V5 to ±18 V or 5 V to 16 V	±2 V to ±18 V or 4 V to 36 V	←——— ±5 V to ±18 V ———→			±5 V to ±18 V
Supply current	1.7 mA	5.5 mA	1.8 mA	3.6 mA	800 µA	1.8 mA	150 µA	2 mA
Input offset voltage	1 mV	2 mV	8 mV	5 mV	5 mV	0.8 mV	1 mV	5 mV
Input bias current	200 nA	400 nA	5 pA	10 pA	50 pA	50 pA	10 pA	50 pA
Input resistance	1 M Ω	20 M	1.5 TΩ	1.5 TΩ	1 TΩ	1 TΩ	1 TΩ	0.5 TΩ
Voltage gain, A_0	106 dB	96 dB	110 dB	100 dB	88 dB	106 dB	100 dB	100 dB
CMRR	90 dB	100 dB	90 dB	90 dB	100 dB	100 dB	95 dB	90 dB
f_T	1 MHz	1 MHz	15 MHz	4.5 MHz	4 MHz	4 MHz	1 MHz	1 MHz
Slew rate	0.5 V/µS	35 V/µS	10 V/µS	9 V/µS	13 V/µS	15 V/µS	1 V/µS	0.5 V/µS
8-pin DIL outline	b	a	c	c	b	b	b	b

Figure 1.5　Parameter and outline details of eight popular 'single' op-amp types

Linear amplifier circuits

Figures 1.6 to *1.12* show a variety of ways of using op-amps to make practical linear amplifier circuits. Note that although type 741 op-amps are specified in the diagrams, any of the op-amp types listed in *Figure 1.5* can in fact be used in these circuits.

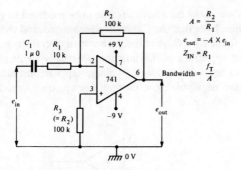

Figure 1.6 *Inverting ac amplifier with × 10 gain*

Figure 1.6 shows the op-amp connected as an inverting ac amplifier with a × 10 overall voltage gain. Note that the pin-3 non-inverting input terminal is tied to ground via R_3, which has the same value as R_2 in order to preserve the dc balance of the op-amp.

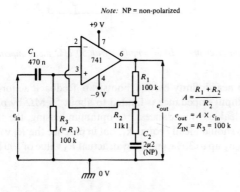

Figure 1.7 *Non-inverting ac amplifier with × 10 gain*

Figure 1.7 shows the op-amp wired as a non-inverting ac amplifier with an overall voltage gain of × 10 $((R_1 + R_2)/R_2))$. Note that R_1 and R_2 are isolated from ground via C_2; at normal operating frequencies C_2 has negligible ac impedance, so ac voltage gain is determined by the R_1 and R_2 ratios, but the op-amp's inverting terminal is subjected to 100% dc negative feedback via R_1, so the circuit has excellent dc stability. For optimum biasing, R_3 has the same value as R_1. The op-amp's input impedance, looking directly into pin-3, is several hundred MΩ, but is shunted by R_3, which reduces the circuit's input impedance to 100 k.

Figure 1.8 shows how the above circuit can be modified to give a 50 MΩ input impedance. Note that the C_2 position is changed, and that the low end of R_3 is taken to the $C_2 - R_2$ junction, rather than directly to ground. The ac feedback signal appearing on this junction is virtually identical to the pin-3 input signal, so near-identical signal voltages appear on both ends of R_3, which thus passes negligible signal current. The apparent impedance of R_3 is

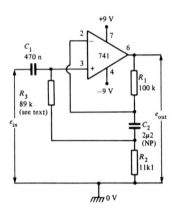

Figure 1.8   *Non-inverting × 10 ac amplifier with 50 MΩ input impedance*

thus raised to near-infinity by this 'bootstrap' feedback action. In practice, this circuit's input impedance is limited to about 50 MΩ by printed circuit board (PCB) leakage impedances. For optimum biasing, the sum of the R_2 and R_3 values should ideally equal R_1, but in practice the R_3 value can differ from the ideal by up to 30%, enabling an actual R_3 value of 100 k to be used if desired.

Voltage follower circuits

An op-amp non-inverting ac amplifier will act as a precision ac voltage follower if wired to give unity voltage gain, and *Figure 1.9* shows some of the design possibilities of such a circuit, which has 100% negative feedback applied from output to input via R_2. Ideally, R_1 (which determines the circuit's input impedance) and R_2 should have equal values, but in practice the R_2 value can be varied from zero to 100 k without greatly upsetting circuit accuracy. If a low f_T op-amp (such as the 741) is used, the R_2 value can usually be reduced to zero. Note, however, that 'high f_T' op-amps tend towards instability when used in the unity-gain mode, and in such cases stability can be

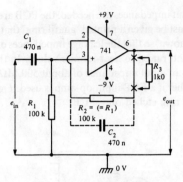

Figure 1.9   *AC voltage follower with 100 k input impedance*

assured by giving R_2 a value of 1k0 or by replaced it with a 1k0 and 100 k resistor wired in series (as shown in *Figure 1.9*), with a 470 n capacitor wired across the 100 k resistor to reduce its ac impedance.

If a very high input impedance is wanted from an ac follower it can be obtained by using the circuit of *Figure 1.10*, in which R_1 is 'bootstrapped' from the op-amp output via C_2, so that the R_1 impedance is increased to near-infinity. In practice, this circuit gives an input impedance of about 50 MΩ from a 741 op-amp, this limit being set by the leakage impedances of the op-amp's IC socket and the PCB.

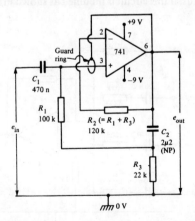

Figure 1.10   *ac voltage follower with 50 MΩ input impedance without the guard ring, or 500 MΩ with the guard ring*

10   Audio processing circuits

If even greater input impedances are needed, the PCB area surrounding the op-amp input pin must be given a printed 'guard ring' that is driven from the op-amp output, as shown, so that the leakage impedances of the PCB, etc., are also bootstrapped and raised to near-infinite values. In this case the *Figure 1.10* circuit will give an input impedance of about 500 MΩ when using a 741 op-amp, or even higher if a FET-input op-amp is used. *Figure 1.11* shows an example of a guard ring etched on a PCB.

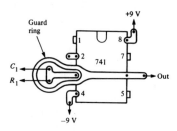

Figure 1.11   *Guard ring etched on a PCB and viewed through the top of the board*

An audio mixer

When describing the basic inverting amplifier circuit of *Figure 1.4(a)* it was pointed out that its voltage gain equals R_2/R_1; the signal flowing in R_1 and R_2 are thus always exactly equal (but are opposite in phase), irrespective of the R_1 and R_2 values. Thus, if this circuit is modified as shown in *Figure 1.12*, where

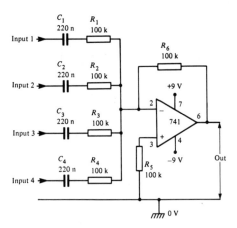

Figure 1.12   *Four-input audio mixer*

four identical input networks are wired in parallel, the feedback signal current flowing in R_6 will inevitably equal the sum of the input signal currents flowing in resistors R_1 to R_4, and the circuit's output signal voltage is thus proportional to the sum of the audio input signal voltages. When (as in the diagram) the input and feedback resistors have equal values, this circuit gives unity voltage gain between each input and the output.

It can thus be seen that this circuit actually functions as a unity-gain 4-input (or 4-channel) audio mixer that gives an output equal to the sum of the four input signal voltages. If desired, this simple circuit can be converted into a practical audio mixer by feeding each input signal to its input network via a 10 k 'volume control' pot. If desired, the circuit can be made to give a voltage gain greater than unity by increasing the R_6 value; the number of available input channels can be increased (or reduced) by adding (or deleting) one new $C_1 - R_1$ network for each new channel.

Active filters

Filter circuits are used to reject unwanted frequencies and pass only those wanted by the designer. A simple R–C low-pass filter (*Figure 1.13(a)*), passes low-frequency signals, but rejects high-frequency ones. The output falls by 3 dB at a 'break' or 'crossover' frequency (f_c) of $1/(2\pi RC)$, and then falls at a rate of 6 dB/octave (= 20 dB/decade) as the frequency is increased (see *Figure*

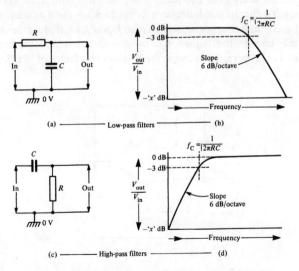

Figure 1.13   *Circuits and response curves of simple 1st-order* R–C *filters*

1.13(b)). Thus, a 1 kHz filter gives about 12 dB of rejection to a 4 kHz signal, and 20 dB to a 10 kHz one.

A simple *R–C* high-pass filter (*Figure 1.13(c)*) passes high-frequency signals but rejects low-frequency ones. The output is 3 dB down at a break frequency of $1/(2\pi RC)$, and falls at a 6 dB/octave rate as the frequency is decreased below this value (see *Figure 1.13(d)*). Thus, a 1 kHz filter gives 12 dB of rejection to a 250 Hz signal, and 20 dB to 100 Hz.

Each of the above two filter circuits uses a single *R–C* stage, and is known as a '1st order' filter. If we could simply cascade a number (*n*) of these filter stages, the resulting circuit would be known as an '*n*th order' filter and would have an output slope, beyond f_c, of $(n \times 6\ \text{dB})$/octave. Thus, a 4th order 1 kHz low-pass filter would have a slope of 24 dB/octave, and would give 48 dB of rejection to a 4 kHz signal, and 80 dB to a 10 kHz signal.

Unfortunately, simple *R–C* filters cannot be directly cascaded, since they would then interact and give poor results; they can, however, be *effectively* cascaded by incorporating them into the feedback networks of suitable op-amp circuits. Such circuits are known as 'active filters', and *Figures 1.14* to *1.20* show practical examples of some of them.

Active filter circuits

Figure 1.14 shows the practical circuit and formula of a maximally flat (Butterworth) unity-gain 2nd-order low-pass filter with a 10 kHz break frequency. To alter its break frequency, change either the *R* or the *C* value in proportion to the frequency ratio relative to *Figure 1.14*. Reduce the values by this ratio to increase the frequency, or increase them to reduce the frequency. Thus, for 4 kHz operation, increase the *R* values by a factor of 10 kHz/4 kHz, or 2.5 times.

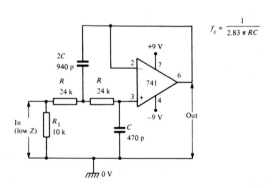

$$f_c = \frac{1}{2.83\,\pi\,RC}$$

Figure 1.14   *Unity-gain 2nd-order low-pass active filter*

A minor snag with the *Figure 1.14* circuit is that one of its 'C' values should ideally be precisely twice the value of the other, and this can result in some rather odd component values. *Figure 1.15* shows an alternative 2nd-order 10 kHz low-pass filter circuit that overcomes this snag and uses equal component values. Note here that the op-amp is designed to give a voltage gain of 4.1 dB via R_1 and R_2, which must have the values shown.

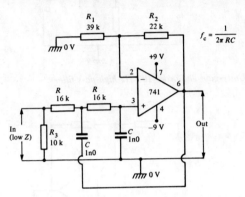

Figure 1.15   *'Equal components' version of 2nd-order 10 kHz low-pass active filter*

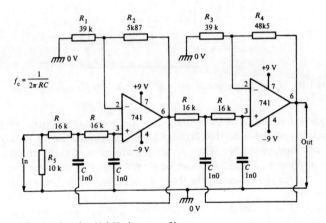

Figure 1.16   *A 4th-order 10 kHz low-pass filter*

Figure 1.16 shows how two of these 'equal component' filters can be cascaded to make a 4th-order low-pass filter with a slope of 24 dB/octave. In this case gain-determining resistors R_1/R_2 have a ratio of 6.644, and R_3/R_4 have a ratio of 0.805, giving an overall voltage gain of 8.3 dB. The odd values of R_2 and R_4 can be made by series-connecting standard 5% resistors.

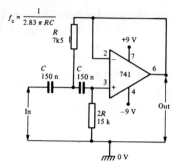

Figure 1.17   *Unity-gain 2nd-order 100 Hz high-pass filter*

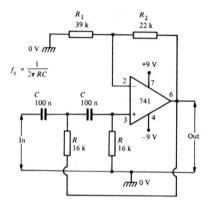

Figure 1.18   *'Equal components' version of 2nd-order 100 Hz high-pass filter*

Figures 1.17 and *1.18* show unity-gain and 'equal component' versions respectively of 2nd-order 100 Hz high-pass filters, and *Figure 1.19* shows a 4th-order high-pass filter. The operating frequencies of these circuits, and those of *Figures 1.15* and *1.16*, can be altered in exactly the same way as in *Figure 1.14*, i.e., by increasing the R or C values to reduce the break frequency, or vice versa.

Finally, *Figure 1.20* shows how the *Figure 1.18* high-pass and *Figure 1.15* low-pass filters can be wired in series to make (with suitable component value changes) a 300 Hz to 3.4 kHz speech filter that gives 12 dB/octave rejection to all signals outside of this range. In the case of the high-pass filter, the 'C' values of *Figure 1.18* are reduced by a factor of three, to raise the break frequency from 100 Hz to 300 Hz, and in the case of the low-pass filter the 'R' values of *Figure 1.15* are increased by a factor of 2.94, to reduce the break frequency from 10 kHz to 3.4 kHz.

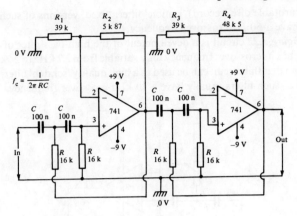

Figure 1.19 *A 4th-order 100 Hz high-pass filter*

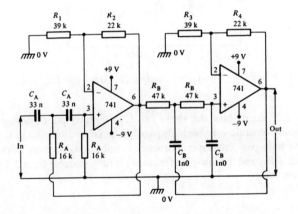

Figure 1.20 *300 Hz to 3.4 kHz speech filter with 2nd-order response*

Variable active filters

The most useful type of active filter is that in which the crossover frequency is fully and easily variable over a fairly wide range, and *Figures 1.21* to *1.23* show three practical examples of 2nd-order versions of such circuits.

The *Figure 1.21* circuit is a simple development of the high-pass filter of *Figure 1.17*, but has its crossover frequency fully variable from 23.5 Hz to 700 Hz via RV_1. The reader should note that this circuit can be used as a high quality turntable disc (record) 'rumble' filter; 'fixed' versions of such filters usually have a 50 Hz crossover frequency.

The *Figure 1.22* circuit is a development of the high-pass filter of *Figure 1.14*, but has its crossover frequency fully variable from 2.2 kHz to 24 kHz via RV_1. Note that this circuit can be used as a high quality 'scratch' filter; 'fixed' versions of such filters usually have a 10 kHz crossover frequency.

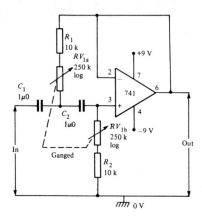

Figure 1.21　*Variable high-pass filter, covering 23.5 Hz to 700 Hz*

Figure 1.23 shows how the above two filter circuits can be combined to make a really versatile variable high-pass/low-pass or rumble/scratch/speech filter. The high-pass crossover frequency is fully variable from 23.5 Hz to 700 Hz via RV_1, and the low-pass frequency is fully variable from 2.2 kHz to 24 kHz via RV_2.

Tone-control networks

The most widely used types of variable filter circuit are those used in audio tone-control applications. These allow the user to alter a systems frequency response to either suit his/her individual needs/moods, or to compensate for anomalies in room acoustics, etc. We shall be looking at practical examples of such circuits shortly. First, however, let us look at some basic tone-control networks.

Figure 1.24(a) shows the typical circuit of a passive bass tone control network (which can be used to boost effectively or cut the low-frequency parts of the audio spectrum), and *Figure 1.24(b)* to *(d)* shows the equivalent of this circuit when RV_1 is set to the maximum BOOST, maximum CUT, and FLAT

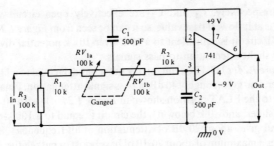

Figure 1.22   *Variable low-pass filter, covering 2.2 kHz to 24 kHz*

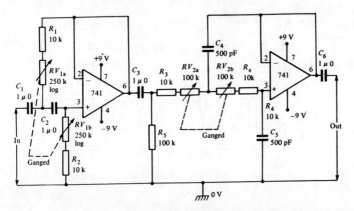

Figure 1.23   *Variable high-pass/low-pass or rumble/scratch/speech filter*

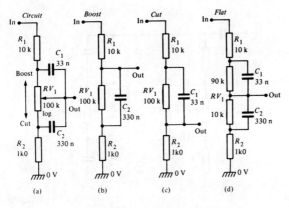

Figure 1.24   *Circuit and equivalents of bass tone control network*

positions respectively. C_1 and C_2 are effectively open circuit when the frequency is at its lowest bass value, so it can be seen from *Figure 1.24(b)* that the BOOST circuit is equivalent to a 10 k-over-101 k potential divider, and gives only slight attenuation to bass signals.

The *Figure 1.24(c)* CUT circuit, on the other hand, is equal to a 110 k-over-10 k divider, and gives roughly 40 dB of bass signal attenuation. Finally, when RV_1 is set to the FLAT position shown in *Figure 1.24(d)* (with 90 k of RV_1 above the slider, and 10 k below it), the circuit is equal to a 100 k-over-11 k divider, and gives about 20 dB of attenuation at all frequencies. Thus, the circuit gives a maximum of about 20 dB of bass boost or cut relative to the flat signals.

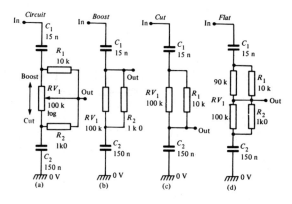

Figure 1.25   *Circuit and equivalents of treble tone control network*

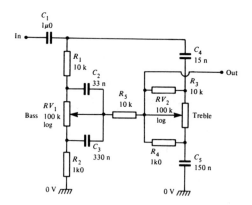

Figure 1.26   *Passive bass and treble tone control network*

Figure 1.25 shows the typical circuit of a passive treble tone control network (which can be used to effectively boost or cut the high-frequency parts of the audio spectrum), together with its equivalent circuits under the maximum boost, maximum cut, and flat operating conditions. This circuit gives about 20 dB of signal attenuation when RV_1 is in the flat position, and gives maximum treble boost or cut values of 20 dB relative to the flat performance.

Figure 1.26 shows how the *Figures 1.24(a)* and *1.25(a)* circuits can be combined to make a complete passive bass and treble tone control network; 10 k resistor R_5 helps minimize unwanted interaction between the two circuit sections. This tone control circuit can be interposed between an amplifier's volume control and the input of its power amplifier stage.

Active tone controls

An active tone control circuit can easily be made by wiring a passive tone control network into the negative feedback loop of an op-amp linear amplifier, so that the system gives an overall signal gain (rather than attenuation) when its controls are in the flat position. Such networks can take the form of simplified versions of the basic *Figure 1.26* circuit, but more often are based on the alternative passive tone control circuit shown in *Figure 1.27*, which gives a similar performance but uses fewer components and uses linear control pots.

Looking at *Figure 1.27*, it can be seen that at very low frequencies (when the two capacitors act like open circuits) the output signal amplitudes are

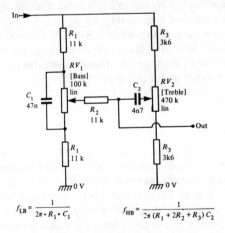

$$f_{\text{LB}} = \frac{1}{2\pi \cdot R_1 \cdot C_1} \qquad f_{\text{HB}} = \frac{1}{2\pi (R_1 + 2R_2 + R_3) C_2}$$

Figure 1.27   *Alternative tone control circuit*

controlled entirely by RV_1 (since RV_2 is isolated from the output via C_2), but that at high frequencies (when the two capacitors act like short circuits) the output signal amplitudes are controlled entirely by RV_2 (since RV_1 is shorted out via C_1). The low-frequency (bass) break point of the circuit is determined by the R_1–C_1 values, and the high-frequency (TREBLE) break point is determined by C_2 and the values of R_1 to R_3.

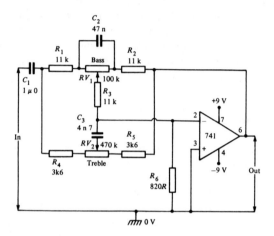

Figure 1.28   *Active tone control circuit*

Figure 1.28 shows how to use the above network to make a practical active tone control circuit that can give up to 20 dB of boost or cut to bass or treble signals. This is an excellent high-quality design.

An even more useful circuit is shown in *Figure 1.29*. This design is similar to the above, but has an additional filter control network that is centred on the 1 kHz 'midband' part of the spectrum, thus enabling this part of the audio band also to be boosted or cut by up to 20 dB.

Graphic equalizers

The ultimate (most sophisticated) type of tone control system is the so-called graphic equalizer. This consists of a number of parallel-connected, overlapping, narrow-band, variable-response filters that cover the entire audio spectrum, thus enabling an amplifier system's spectral response to be precisely adjusted to suit individual needs. Usually, the filter centre frequencies are spaced at one octave intervals and such systems are thus also known as octave equalizers.

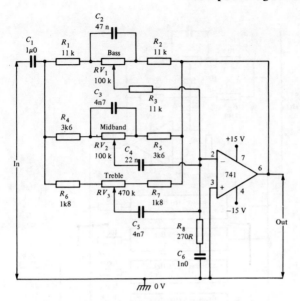

Figure 1.29 *Three-band (bass, midband, treble) active tone control circuit*

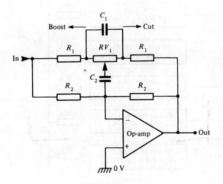

Figure 1.30 *Typical octave (graphic) equalizer section*

Figure 1.30 shows the basic circuit of a typical octave (graphic) equalizer section. This circuit is in fact very similar to that of the *Figure 1.28* active tone control, except that the C_2–R_2 'treble control' network is fixed, rather than variable, and the bass and treble break frequencies are fairly closely spaced, so that the two response curves overlap. The net effect of this is that the *Figure 1.30* circuit acts as a narrow-band filter which has a centre-frequency response

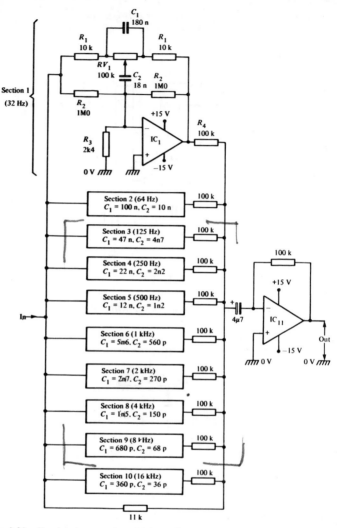

Figure 1.31 *Ten band octave (graphic) equalizer circuit*

that is fully adjustable between plus 12 dB (full boost) and minus 12 dB (full cut) via RV_1.

Figure 1.31 shows how ten of the above circuits can be interconnected to make a practical high quality, ten-band graphic equalizer; the ten equalizer sections are wired in parallel, and their outputs are added together in the IC_{11}

output stage. The ICs used here can be type-741 'single' or 'quad' op-amps. Note that two complete *Figure 1.31* circuits are needed in a normal stereo amplifier system.

Record Industry Association of America equalizers

Three types of phonograph record (disc) pickup are in general use, these being the ceramic, crystal and magnetic types. The first two of these are fairly cheap types which give large amplitude outputs, have a reasonably linear frequency response, and are widely used in low-fi to mid-fi equipment. The third (magnetic) type, on the other hand, gives a low amplitude output and has a non-linear frequency response, but is widely used on good quality hi-fi equipment.

If the reader were to take a test disc (phonograph record) on which a constant amplitude 20 Hz to 20 kHz three-decade span of sine wave tone signals had been recorded with perfect linearity (constant signal amplitude in the recording grooves), and were to then play that disc through a good quality magnetic pickup, he/she would find that it generates a non-linear frequency response that rises at a rate of 6 dB per octave (20 dB per decade). Thus, output signals would be very weak at 20 Hz, but would be 1000 times greater (+60 dB) at 20 kHz.

This non-linear frequency response is inherent in all magnetic pickups, since their output voltage is directly proportional to the rate of movement of the pickup needle, which in turn is proportional to the recording frequency.

In practice, disc recording equipment does not give an exactly linear frequency response. To help enhance the effective dynamic range and signal-to-noise ratio performance of discs, frequencies below 50 Hz and those in the 500 Hz to 2.12 kHz midband range are recorded in a non-linear fashion that is precisely defined by the Record Industry Association of America (RIAA) standards. This non-linearity is such that it causes a midband drop of 12 dB when played through linear-response ceramic or crystal pickups, but this modest decrease is too small to be objectionable in most low-fi to mid-fi playback equipment.

When a practical RIAA test frequency disc is played through a magnetic pickup, the pickup produces the frequency response curve shown in *Figure 1.32*. Here, the dotted line shows the idealized shape of this curve, which is flat up to 50 Hz, then rises at a 6 dB/octave rate to 500 Hz, then is flat to 2120 Hz, and then rises at a 6 dB/octave rate beyond that. The solid line shows the practical shape of the curve.

The really important point that the reader should note from all this is that when a disc is played through a magnetic pickup in a good quality hi-fi system, the output of the pickup must be passed to the power amplifier circuitry via a

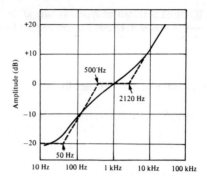

Figure 1.32   *Typical phono disc playback frequency response curve*

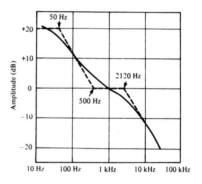

Figure 1.33   *RIAA playback equalization curve*

pre-amplifier that has a frequency equalization curve that is the exact inverse of that shown in *Figure 1.32*, so that a linear overall record-to-replay response is obtained. *Figure 1.33* shows the actual form of the necessary RIAA equalization curve, and *Figure 1.34* shows a practical example of a modern low-noise phono pre-amplifier with integral RIAA magnetic-pickup equalization.

RIAA phono pre-amp

Magnetic pickups are low-sensitivity devices, and give typical midband outputs of only a few millivolts. Consequently, their outputs must be passed to main amplifiers via dedicated low-noise pre-amplifier ICs (rather than via simple op-amps), and the *Figure 1.34* circuit is thus designed around LM381 or LM387 ICs of this type. These are 'dual' ICs, and in the diagram the pin numbers of one half of the IC are shown in plain numbers, and the other half are shown in brackets. Note that two of these pre-amp circuits are needed in a stereo audio system, and these can thus be obtained from one 'dual' IC.

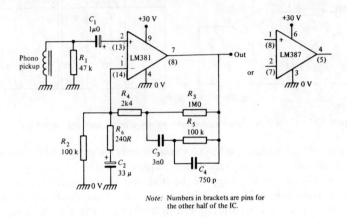

Note: Numbers in brackets are pins for
the other half of the IC.

Figure 1.34   *LM381 (or LM387) low-noise phono pre-amp (RIAA)*

The operating theory of the *Figure 1.34* circuit is fairly simple. The IC is wired as a non-inverting amplifier, with negative feedback applied from the output to the inverting input terminal; potential divider $(R_3 + R_4) - R_2$ determines the circuit's dc biasing, and the $R_5-C_3-C_4-R_4$ and R_5-C_2 networks determine the ac signal gain. At the 1 kHz midband frequency, C_2 and C_3 have low impedances and C_4 has a high impedance, so the ac gain is determined mainly by R_5/R_5, and equals $\times 400$. At lower frequencies the impedance of C_3 starts to become significant and causes the ac gain to increase until eventually, at very low frequencies, it is limited to $\times 4000$ by the R_3/R_5 ratio. At high frequencies, on the other hand, the impedance of C_4 falls to significant levels and shunts R_5, thus causing the ac gain to decrease until eventually, at very high frequencies, it is limited to $\times 10$ by the R_4/R_5 ratio. The pickup signals are ac coupled to the IC via C_1, and the circuit can be used with any type of magnetic pickup unit.

Non-linear amplifiers

An op-amp circuit can be made to act as a non-linear amplifier by simply incorporating a non-linear device in its negative feedback network, as shown in *Figure 1.35*, where the feedback elements comprise a pair of silicon diodes

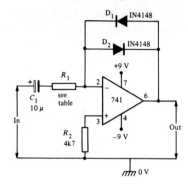

V_{in} (rms)	R_1 = 1k0		R_1 = 10 k	
	V_{out} (mV rms)	V_{gain}	V_{out} (mV rms)	V_{gain}
1 mV	110	× 110	21	× 21
10 mV	330	× 33	170	× 17
100 mV	450	× 4.5	360	× 3.6
1 V	560	× 0.56	470	× 0.47
10 V	600	× 0.07	560	× 0.056

Figure 1.35   *Circuit and performance table of non-linear (semi-log) amplifier*

connected back-to-back. When very small signals are applied to this circuit the diodes act like very high resistances, so the circuit gives high voltage gain, but when large signals are applied the diodes act like low resistances, so the circuit gives low gain. The gain in fact varies in a semilogarithmic fashion, and circuit sensitivity can be varied by altering the R_1 value; the table shows actual performance details. Note that a 1000:1 change in input signal amplitude can cause as little as a 2:1 change in output level, so, by taking its output to a simple AC millivoltmeter, this circuit can be used as a single-range bridge-balance detector or signal strength indicator, etc.

When this circuit is fed with a sine wave input its two diodes clip the output to about 1.4 V peak-to-peak, giving an almost square output that is rich in

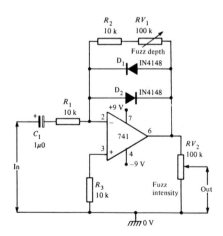

Figure 1.36   *'Fuzz' circuit*

odd harmonics, and which, when fed through an amplifier, sounds like a clarinet; in the music world this is known as a 'fuzz' effect. *Figure 1.36* shows how the circuit can be modified to make a practical 'fuzz' generator; RV_1 controls the level at which fuzz clipping begins, and RV_2 controls the circuit's output level or fuzz intensity.

Constant-volume amplifier

The non-linear amplifier of *Figure 1.35* gives a near constant-amplitude output signal over a wide range of input levels, but does so by causing heavy amplitude distortion of the signal. *Figure 1.37* shows an alternative type of 'constant volume' amplifier in which amplitude control is obtained *without* generating signal distortion; this is achieved by using a voltage-controlled linear element (rather than a non-linear element) in its negative feedback loop.

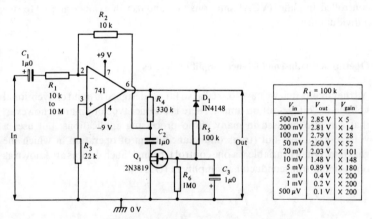

R_1 = 100 k		
V_{in}	V_{out}	V_{gain}
500 mV	2.85 V	X 5
200 mV	2.81 V	X 14
100 mV	2.79 V	X 28
50 mV	2.60 V	X 52
20 mV	2.03 V	X 101
10 mV	1.48 V	X 148
5 mV	0.89 V	X 180
2 mV	0.4 V	X 200
1 mV	0.2 V	X 200
500 μV	0.1 V	X 200

Figure 1.37 *Circuit and performance table of constant-volume amplifier*

Here, the op-amp is wired as an ac amplifier with its gain controlled by the R_2/R_1 ratio and by the ac potential divider formed by R_4 and the drain impedance of field-effect transistor Q_1, which is used as a voltage-controlled resistor with its control voltage derived from the op-amp output via the $D_1-R_5-R_5-C_3$ network, which generates a voltage proportional to the mean value (integrated over hundreds of milliseconds) of the output signal. With zero bias applied to Q_1 gate the FET acts like a low resistance (a few hundred ohms), but with a large negative bias voltage applied it acts like a high resistance (a few megohms).

Thus, when a very small input signal is applied the op-amp's output also tends to be small, so D_1–R_5–C_3 feed near-zero negative bias to the gate of the FET, which thus acts like a resistance of only a few hundred ohms. Under this condition the R_4–Q_1 divider causes very little negative feedback to be applied to the op-amp, which thus gives a high voltage gain.

When a large input signal is applied to the op-amp its output tends to be large, so a large negative bias is developed on the gate of the FET, which thus acts like a very high resistance and causes heavy negative feedback to be applied to the op-amp, which thus gives very low voltage gain.

The net effect of this action is that the mean level of the output tends to self-regulate at 1.5 to 2.85 V over a 50:1 (500 mV to 10 mV) range of input signal levels, and does so without generating significant distortion. The R_1 value determines the sensitivity of the circuit, and is selected to suit the maximum input signal amplitude that the circuit is expected to handle, on the basis of 200 k per rms volt of input signal; thus, for a maximum input of 50 V R_1 has a value of 10 M, and for 50 mV it has a value of 10 k. C_3 determines the voltage-controlled amplifier (VCA) time constant, and its value can be altered to suit individual needs.

Operational transconductance amplifier devices

All the circuits that we have looked at so far are designed around conventional voltage-in voltage-out op-amps. There is another type of op-amp, however, that can also be used in many audio processing applications, but uses a voltage-in current-out (transconductance) form of operation in which the gain is externally variable via one control terminal. Such devices are known as operational transconductance amplifiers (OTAs).

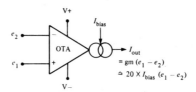

Figure 1.38   *Symbol and basic formulae of a conventional OTA*

Figure 1.38 shows the standard circuit symbol and basic operating formula of the OTA. The device has conventional differential voltage input terminals, which accept inputs e_1 and e_2, and gives an output current that equals the difference between these signals multiplied by the OTA's transconductance or 'gm' value, which in turn typically equals 20 times the external bias current value. Thus, the gain can be controlled by the bias current which, in practice, can easily be varied over a 10,000:1 range.

OTAs are reasonably versatile devices. They can be made to act like voltage-in voltage-out op-amps by simply feeding their output current into a load resistor, which converts the current into a voltage, and their gain can be voltage-controlled by applying the voltage via a series resistor, which converts it into a control current. By using these techniques, the OTA can be used as a VCA, as an amplitude modulator, or as a ring modulator or four-quadrant multiplier, etc.

Practical OTAs

The two best-known versions of the OTA are the CA3080 and the LM13600. The CA3080 is a simple first-generation device that can accept bias currents in the range 100 nA to 1 mA and can operate from split power supplies in the 2 to 15 V range; the device is housed in an 8-pin DIL package with the outline and pin notations shown in *Figure 1.39*. A minor defect of this IC is that it generates a certain amount of signal distortion.

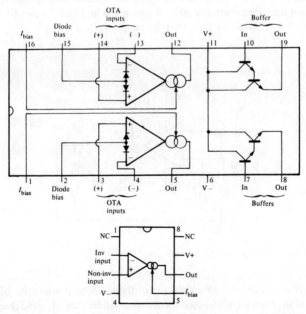

Figure 1.39   *Outline and pin notations of the CA3080 (lower) and LM13600 (upper) OTAs*

The LM13600 is an improved second-generation version of the OTA, and incorporates linearizing input diodes that greatly reduce signal distortion, plus a coupled output buffer stage that can be used to give a low-impedance output. The LM13600 is actually a dual OTA, and *Figure 1.39* also shows the circuit symbol, IC outline, and pin notations of this device, which is housed in a 16-pin DIL package.

Basic OTA circuits

The CA3080 and LM13600 OTAs are very easy ICs to use, and in this section we show how to use them in basic fixed-gain ac amplifier applications. Looking first at the CA3080, its pin-5 I_{bias} terminal is actually connected to the pin-4 negative supply rail via an internal base-emitter junction, so the biased voltage of pin-5 is about 600 mV above the pin-4 voltage. I_{bias} can thus be obtained by simply connecting pin-5 to either the common rail or the positive supply rail via a current-limiting resistor of suitable value.

Figure 1.40 shows a simple but instructive way of using the CA3080 as an ac-coupled inverting amplifier with a voltage gain of about 40 dB. The circuit is operated from split 9V supplies, so 17.4 V are generated across bias resistor R_3, which thus feeds roughly 500 μA into pin-5 and thus causes the IC to consume 1 mA (20 times I_{bias}) from its supply rails.

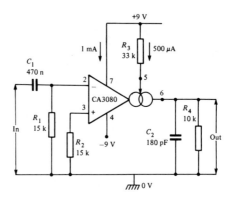

Figure 1.40   *ac-coupled 40 dB inverting amplifier*

At a bias current of 500 μA the gm of the CA3080 is about 10 mS. The output of the *Figure 1.40* circuit is loaded by a 10 k resistor (R_4), and thus gives an overall voltage gain of 10 mS × 10 k = × 100, or 40 dB. The peak current that can flow into this 10 k load is 500 μA ($= I_{bias}$), so the peak available output voltage is plus or minus 5 V. The output is also loaded by 180 pF capacitor C_2, which limits the output slew rate to about 2.8 V/μs.

Each of the two OTA devices housed in the LM13600 package can, if desired, be used in exactly the same way as the CA3080 shown in *Figure 1.40*. One of the main features of the LM13600, however, is that it incorporates linearizing diodes that help reduce signal distortion, so in practice each half of the device is best used as an inverting ac amplifier by wiring it as shown in *Figure 1.41*, where the two input diodes are biased, via R_3, with current I_D, which flows to ground via R_2 and R_4.

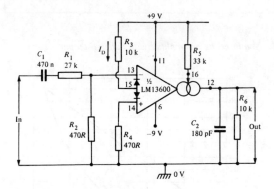

Figure 1.41   *Inverting ac amplifier with near-unity overall voltage gain*

Note in this circuit that the input signal is applied to the non-inverting input terminal via R_1, and that R_1 and R_2 form a voltage divider that attenuates the input signal, and that the circuit consequently gives an overall voltage gain of slightly less than unity. The gain of this circuit is in fact proportional to the value of I_{bias}/I_D, and can thus be varied by altering the value of either I_{bias} or I_D.

The above circuit can be made to act as a non-inverting amplifier by modifying the input circuitry as shown in *Figure 1.42*, which also shows how it can be made to give a low impedance output by feeding the OTA output to the outside world via one of the IC's internal buffer amplifiers; this modification enables the R_6 value to be increased to 33 k, with a consequent increase in overall voltage gain.

The graph of *Figure 1.43* shows the typical signal distortion figures obtained from the LM13600 when used with and without the internal linearizing diodes. With an input terminal signal of 20 mV peak-to-peak the device generates less than 0.02% distortion with the diodes, but about 0.3% without them; these figures rise to 0.035% and 1.5% respectively at 40 mV input.

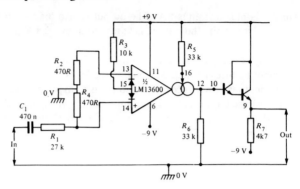

Figure 1.42   *Non-inverting ac amplifier with buffered output*

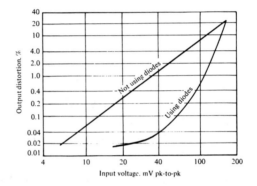

Figure 1.43   *Typical distortion levels of the LM13600 OTA, with and without the use of the linearizing diodes*

CA3080 variable-gain circuits

Figure 1.44 shows how the basic *Figure 1.40* inverting amplifier circuit can be modified so that its gain is variable from × 5 to × 100 via RV_2, which enables I_{bias} to be varied from 12.4 μA to 527 μA. In this type of application the input bias levels of the IC must be balanced so that the output dc level does not shift as the gain is varied, and this is achieved via offset-nulling preset pot RV_1. To set up the circuit, set RV_2 to its minimum (maximum gain) value, and then trim RV_1 to give zero dc output.

    The *Figure 1.44* circuit can be converted into a voltage-controlled amplifier by removing R_4 and RV_2 and connecting the V_{in} voltage-control input to pin-

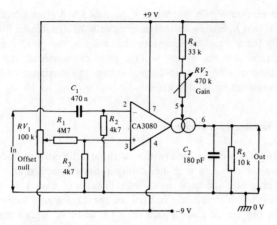

Figure 1.44   *Variable gain (× 5 to × 100) ac amplifier*

5 of the CA3080 via a 33 k series resistor. In this case the circuit gives a gain of × 100 when V_{in} equals the positive supply rail value, and near-zero gain when V_{in} is 600 mV above the negative supply rail value. Thus, to give the full range of gain control, V_{in} must be referenced to the negative supply rail.

Figure 1.45 shows a more useful type of VCA, in which V_{in} is referenced to the common (zero) supply rail. Here, Q_1 and the 741 op-amp form a linear voltage-to-current converter (with a 100 μA/V conversion rate) which responds to positive V_{in} values only. Thus, when V_{in} equals zero or less, the

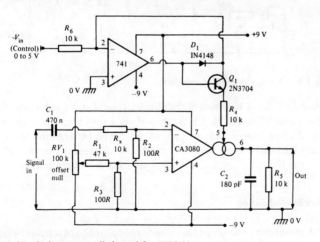

Figure 1.45   *Voltage controlled amplifier (VCA)*

VCA gives near-zero gain, but when V_{in} equals 5 V it gives a basic gain of × 100. Note that R_x is shown wired in series with the signal input line; R_x and R_2 actually form a potential divider that reduces the pin-2 input signal amplitude (and thus the overall voltage gain) of the circuit; the R_x value should be chosen to limit the pin-2 signal voltage to a maximum of 20 mV pk-to-pk, to minimize signal distortion.

LM13600 variable-gain circuits

The LM13600 OTA can be used (with or without its linearizing diodes) in any of the basic variable-gain amplifier configurations described above. *Figure 1.46*, for example, shows it used with linearizing diodes in the VCA configuration in which the gain control voltage is referenced to the negative supply rail; this circuit gives a gain of × 1.5 when V_{in} equals the positive supply rail value, and gives an attenuation of 80 dB when V_{in} equals the negative supply rail value.

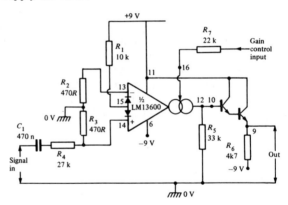

Figure 1.46   *Voltage controlled amplifier*

Figure 1.47 shows how two of the above circuits can be joined together to make a stereo VCA unit that is controlled via a single input voltage, which may be derived from a 'volume control' pot wired between the two supply rails, in which case a 10 μF capacitor can be wired across the lower half of the pot so that the circuit acts as a 'noiseless' volume control system.

Amplitude modulation

A VCA circuit can be used as an amplitude modulator (AM) circuit by feeding a carrier signal to its input terminal and using a modulating signal to control

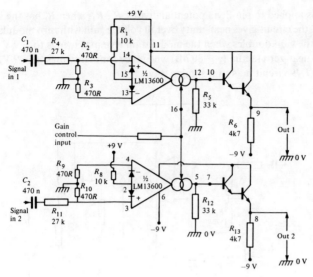

Figure 1.47   *Stereo VCA*

the output amplitude via the gain-control input terminal. *Figure 1.48* shows a
CA3080 used in a dedicated version of such a circuit, and *Figure 1.49* shows an
LM13600 version of the same basic design.

The *Figure 1.48* circuit acts as an inverting amplifier; its dc gain is set via
R_4 and R_6, but its ac gain is variable via signals applied to C_2. Input bias
resistors R_1 and R_2 have low values to minimize the IC's noise levels and
enhance stability; offset biasing is applied via R_3–RV_1. The carrier input

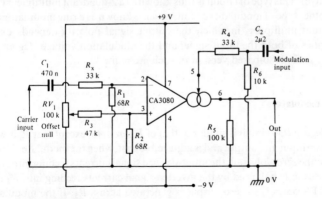

Figure 1.48   *Amplitude modulator (1)*

signal is applied to pin-2 via potential divider R_x–R_1; when R_x has the value shown, the circuit gives near-unity overall voltage gain with zero modulation input; the gain doubles when the modulation terminal swings to +9 V, and falls to near-zero (actually −80 dB) when the terminal swings to −9 V. The *Figure 1.49* circuit acts in a similar way.

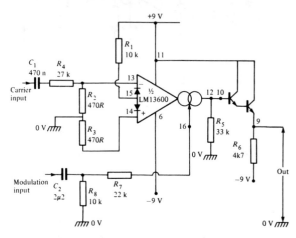

Figure 1.49   *Amplitude modulator (2)*

Note in the above two circuits that the instantaneous polarity of the output signals is determined entirely by the instantaneous polarity of the carrier input signal, which has two possible states (positive or negative), and is independent of the modulation signal, which has only one possible state (positive). This type of circuit is thus known as a 2-quadrant multiplier. There is another type of modulator circuit, which is known as a ring modulator or 4-quadrant multiplier, in which the output signal polarity depends on the polarities of both the input signal and the modulation voltage. *Figure 1.50* shows a CA3080-based version of such a circuit.

Ring modulators

The *Figure 1.50* circuit is similar to that of *Figure 1.48*, except that R_y is wired between input and output and is adjusted so that, when the modulator input is tied to the zero-volts rail, the input-derived signal currents feeding into R_5 via R_y are exactly balanced by the inverted signal currents feeding into R_5 from the OTA output, so zero output is generated across R_5. If the modulation input then goes +ve, the OTA output current exceeds that of the R_y network,

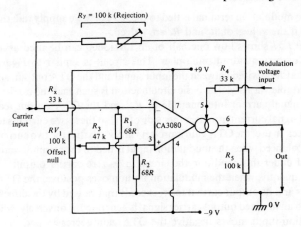

Figure 1.50   *Ring modulator or 4-quadrant multiplier*

and an inverted gain-controlled output is obtained, but if the modulation
ir.put goes $-$ve the R_y output current exceeds that of the OTA, and a non-
inverted gain-controlled output is obtained. Thus, both the phase and the
amplitude of the output signal of this 4-quadrant multiplier are controlled by
the modulation signal. The circuit can be used as a ring modulator by feeding
independent ac signals to the two inputs, or as a frequency doubler by feeding
identical sine wave signals to the two inputs.

With the R_x and R_y values shown this circuit gives a voltage gain of $\times 0.5$

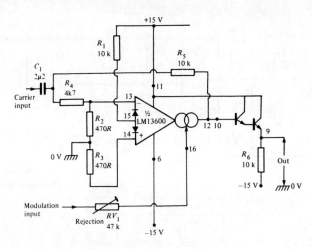

Figure 1.51   *LM13600 ring modulator*

when the modulation terminal is tied to the +ve or −ve supply rail; the gain doubles if the values of R_x and R_y are halved.

Figure 1.51 shows how one half of an LM13600 can be used as a ring-modulator or 4-quadrant multiplier. This circuit is similar to *Figure 1.49* except that R_5 is wired between the input signal and the OTA output, and I_{bias} is pre-settable via RV_1. The basic circuit action is such that the carrier input feeds a signal current into one side of R_5, and the OTA output feeds an inverted signal current into the other side of R_5, so these two currents tend to self-cancel. In use, the OTA gain is pre-set via RV_1 so that the two currents are exactly balanced when the modulation input is tied to the common zero volts line, and under this condition the circuit gives zero carrier output.

Consequently, when the modulation input moves positive, the OTA gain increases and its output current to R_5 exceeds that caused by the direct input signal, so an inverted output carrier signal is generated. Conversely, when the modulation input moves negative the OTA gain decreases and the direct signal current of R_5 exceeds that generated by the OTA output, so a no-inverted output signal is generated.

An automatic gain control amplifier

The gain of the LM13600 OTA can be varied by altering either its I_{bias} or I_D current. *Figure 1.52* shows how I_D variation can be used to make an automatic gain control (AGC) amplifier in which a 100:1 change in input signal amplitude causes only a 5:1 change in output amplitude.

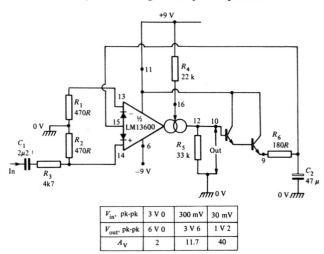

V_{in}, pk-pk	3 V 0	300 mV	30 mV
V_{out}, pk-pk	6 V 0	3 V 6	1 V 2
A_V	2	11.7	40

Figure 1.52   *Circuit and performance table of an AGC amplifier*

In this circuit, I_{bias} is fixed by R_4, and the output signal is taken directly from the OTA via R_5. The output buffer and R_6–C_2 are used to rectify and smooth the OTA output and then to apply an I_D current to the OTA's linearizing diodes. No I_D current is generated until the OTA output exceeds the 1.8 V peak (equals three base-emitter volt drops) needed to turn on the Darlington buffer and the linearizing diodes, but any increase in I_D then reduces the OTA gain and, by negative feedback action, tends to hold V_{out} constant at that level.

The basic gain of this amplifier, with zero I_D, is × 40. Thus, with an input signal of 30 mV pk–pk, the OTA output of 1V2 pk–pk is not enough to generate an I_D current, and the OTA operates at full gain. At 300 mV input, however, the OTA output is enough to generate significant I_D current, and the circuit's negative feedback automatically reduces the output level to 3V6 pk–pk, giving an overall gain of × 11.7. With an input of 3V0, the gain falls to × 2, giving an output of 6 V pk–pk. The circuit thus gives 20:1 signal compression over this range.

The LM13700

The LM13700 is a dual OTA IC that is identical to the LM13600 except for relatively minor differences in its output buffer stages. It is sometimes more readily available than the LM13600, however, and can be used as a pin-for-pin replacement for it in all LM13600 circuits shown in this chapter.

The MC3340P

The MC3340P is very popular and simple dedicated 'electronic attenuator' ICs. *Figure 1.53* shows the outline, pin notations and basic details of the device, which is housed in an 8-pin DIL package; only six of these pins perform useful functions, and two of these are used for power supply connections. Of the remainder, pins 1 and 7 provide input and output signal

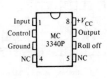

Parameter	Min.	Typ.	Max.
Supply volts	+9 V		+18 V
Control pin sink current			2 mA
Input voltage, rms			0.5 V
Voltage gain		13 dB	
Attenuation range		90 dB	
Total harmonic distortion		0.6%	

Figure 1.53   *Outline and main characteristics of the MC3340P IC*

connections, pin 6 controls roll-off of the device's frequency response, and pin 2 is the device's gain-control terminal.

The basic action of the MC3340P is such that it acts as a simple linear amplifier with 13 dB of signal gain when its pin-2 control terminal is either tied to ground via a 4k0 resistance or is connected to a DC potential of 3.5 V. This gain decreases if the control resistance/voltage is increased above these values, falling by 90 dB (to -77 dB) when the values are increased to 32 k or 6 V. The device's attenuation (or gain) can thus be controlled over a wide range via either a resistance or a voltage.

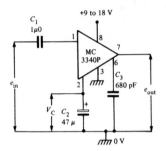
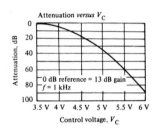

Figure 1.54   *Circuit and performance graph of a voltage-controlled electronic attenuator*

Figure 1.54 shows a practical example of a voltage-controlled MC3340P electronic attenuator, together with its performance graph, and *Figure 1.55* shows a resistance-controlled version of the device. Note in both of these circuits that large-value capacitor C_2 is wired to the control terminal; this helps eliminate control noise and transients, thus giving a 'noiseless' form of gain control and enabling the control resistance/voltage to be remotely located.

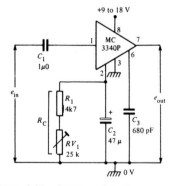
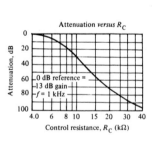

Figure 1.55   *Circuit and performance graph of a resistance-controlled electronic attenuator*

Also note in these circuits that 680 pF capacitor C_3 is wired to pin-6 of the IC; this limits the upper frequency response of the circuit to the high audio range. Without C_3, the response extends to several megahertz and the circuit tends to be unstable. Finally, note that this IC gives very little signal distortion at low attenuation levels, but that distortion rises to about 3% at maximum attenuation values.

The NE570/571 IC

The NE570 is known as a dual 'compander' but is really a rather sophisticated dual VCA IC. Each half (channel) of the IC contains an identical circuit, comprising a full wave rectifier, a variable gain block, an op-amp, a precision 1.8 V reference, and a resistor network. These elements can be externally configured so that each channel acts as either a normal VCA or as a precision dynamic range compressor or expander (hence the compander title).

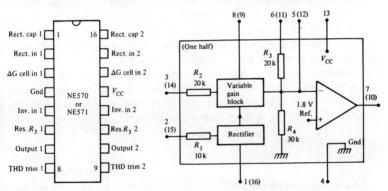

Figure 1.56 *Outline, pin notations and block diagram of the NE570/571 dual compander IC*

Parameter	NE570	NE571
Supply voltage range	6 V to 24 V	6 v to 18 V
Supply current	3.2 mA	3.2 mA
Output current capability	±20 mA	±20 mA
Output slew rate	0.5 V/μS	0.5 V/μS
Gain block distortion:		
Untrimmed	0.3%	0.5%
Trimmed	0.05%	0.1%
Internal reference voltage	1.8 V	1.8 V
Output DC shift	±20 mV	±30 mV
Expander output noise	20 μV	20 μV

Figure 1.57 *Basic characteristics of the NE570 and NE571 ICs*

The NE571 is identical to the NE570, but has a slightly relaxed specification. Both ICs are housed in 16-pin DIL packages. *Figure 1.56* shows the outline and pin designations of the package, together with the block diagram of one IC channel, and *Figure 1.57* lists the basic characteristics of the two ICs. Note in the block diagram (and in following circuits) that pin numbers relating to the left-hand channel of the IC are shown in plain numbers, and those relating to the right-hand half are shown in bracketed numbers.

Circuit description

The operation of the individual elements of the *Figure 1.56* block diagram are fairly easy to understand. Input signals ac coupled to pin 2 (or 15) are full wave rectified and fed to pin 1 (or 16), where they can be smoothed by an external capacitor to generate a VCA control voltage on this pin.

Input signals ac coupled to pin 3 (or 14) are fed to the input of the variable gain block, which is a precision temperate-compensated VCA with its gain controlled via the pin 1 (or 16) voltage; its output is fed to the inverting input of the IC's op-amp stage. Gain block signal distortion is quite low, and can be minimized by feeding a 'trim' voltage to pin 8 (or 9).

The channel's op-amp is internally compensated and has its non-inverting input tied to a 1.8 V precision reference; the inverting input is connected to the gain block output, is externally available, and is also connected to the R_3–R_4 resistor network. The op-amp output is available at pin 7 (or 10).

A stereo VCA

Figure 1.58 shows how a NE570 or NE571 can be used to make a stereo voltage controlled amplifier/attenuator. Here, the internal rectifier is disabled via C_2, and an 0 to 12 V dc control voltage is fed to pins 1 and 16 via R_6 and C_3, to give direct control of the variable gain block. The output of the block is fed to pin 7 (or 10) via the op-amp, which has its ac and dc gain set at × 2.56 via R_4–R_7 and thus generates a quiescent output of 4.62 V (2.56 × 1.8 V). Both channels of the circuit are identical (the control voltage is fed to pins 1 and 16), and give about 6 dB of gain with a control input of 12 V, or 80 dB of attenuation with a control input of zero volts.

Compander theory

In acoustics, the term 'dynamic range' can be simply described as the difference between the loudest and the quietest sound levels that can be

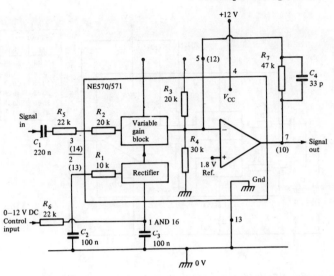

Figure 1.58   *Stereo voltage controlled amplifier/attenuator (only one channel shown)*

perceived or recorded. When listening to music, the human ear has a useful dynamic range of about 90 dB (50,000:1). All practical recording systems generate inherent noise, which limits the minimum strength of signals that can be usefully recorded, and this factor (in conjunction with practical limits on maximum signal strength) places a limit on the useful dynamic range of the recording system.

Simple tape recorder systems typically have a useful dynamic range of only 50 dB, and thus can not directly record and replay high quality music. One way around this problem is to use a compander system to compress the 90 dB dynamic range of the input signal down to 45 dB when recording it (thus giving a 2:1 compression ratio), and then using a matching 1:2 expander to restore its dynamic range to 90 dB when replaying the signals. This same technique can be used to improve the quality of telephone signals, etc., and the NE570/571 ICs are specifically designed for use in these types of application.

An expander circuit

Figure 1.59 shows a practical NE570/571 'expander' circuit and its perform-ance table. Here, the input signal is fed to both the rectifier and the variable gain block, and their action is such that circuit gain is directly proportional to the average value of the input. Thus, if the input rises (or falls) by 6 dB, the

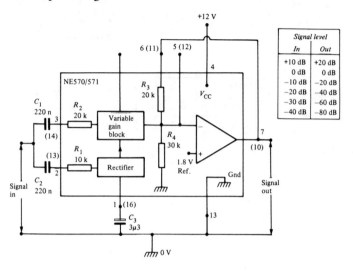

Figure 1.59   *NE570/571 expander circuit and performance table*

gain also rises (or falls) by 6 dB, so the output rises (or falls) by 12 dB, giving a 1:2 expansion ratio. Note in this circuit that (because of the R_3 and R_4 ratios) the op-amp output takes up a quiescent value of 3 V, and can thus supply only modest peak output signals. If desired, the quiescent output can be raised to

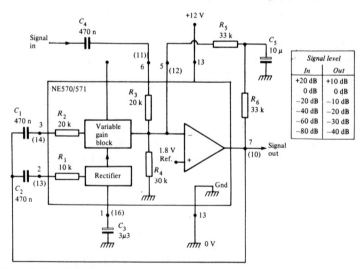

Figure 1.60   *NE570/571 compressor circuit and performance table*

6 V (giving a corresponding increase in peak output levels) by wiring a 12 k resistor in parallel with R_4 via pins 5 (or 12) and 13.

A compressor circuit

Figure 1.60 shows a practical NE570/571 compressor circuit and its performance table. Here, the input signal is fed to the op-amp's inverting input via C_4 and R_3, but the variable gain block and rectifier circuitry are connected in exactly the same way as in the above expander design and are ac coupled into the op-amp's output-to-input negative feedback loop, and the circuit consequently gives a performance that is the exact inverse of the expander, i.e., it gives a 2:1 compression ratio. R_5 and R_6 form a dc feedback loop (ac decoupled via C_5) that biases the op-amp output at a quiescent value of about 6 V.

A total harmonic distortion trimmer

Finally, to complete this chapter, *Figure 1.61* shows a total harmonic distortion (THD) trim network that can be added to the above expander or compressor circuits to minimize their THD figures. To adjust this trimmer, feed a fairly strong 1 kHz sinewave to the input of the compander, and then adjust RV_1 for minimum output distortion.

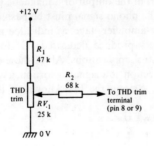

Figure 1.61   *THD trim network*

2 Audio pre-amplifier circuits

Each channel of a modern stereo hi-fi audio system comprises a number of interconnected circuit blocks, as indicated in *Figure 2.1*. Here, input signals from either a radio tuner, a tape (cassette) deck, or a phono pre-amplifier are selected via SW_1 and then fed to the input of a power amplifier stage via a tone-control system and a volume control. In practice, the tone control system may include refinements such as 'scratch' and 'rumble' filters, etc.

Typically, the tone-control system needs to be driven by input signals with mean amplitudes of tens or hundreds of millivolts. Signals of suitable amplitude are usually available directly from the output of a tape or tuner unit, but not directly from the output of a magnetic phono pickup. In the latter case, therefore, the phono signal must be passed to the tone-control input via a suitable pre-amplifier stage, as indicated in the *Figure 2.1*.

Several manufacturers produce dedicated ICs for use in audio pre-amplifier and tone-control applications. All of these devices give excellent power-supply ripple rejection, low signal distortion, a wide bandwidth, and a very low noise figure. Among the best-known of these devices are the

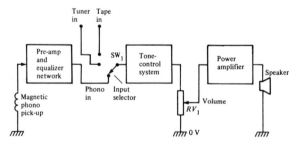

Figure 2.1   *Block diagram of one channel of a hi-fi system*

46

National Semiconductor Corporation's LM381/LM382/LM387 family of dual audio pre-amplifier ICs, and most of this chapter is devoted to taking an in-depth look at this range of devices. We also take a brief look at one other popular (but slightly less well known) pre-amp IC, however, and this is the TDA3410, which is another ultra-low-noise dual pre-amp.

LM381/LM382/LM387 ICs

National Semiconductor Corporation (NSC) produce a range of five low-noise dual pre-amp ICs, these being the LM381 and LM381A, the LM382, and the LM387 and LM387A: the 'A' suffix devices are simply premium versions of their type, with superior low-noise figures. *Figures 2.2* to *2.4* show the outlines of each of these ICs, together with the actual circuit of one of the identical pair of amplifiers that are housed in each package, and *Figure 2.5* gives a summary of their performances.

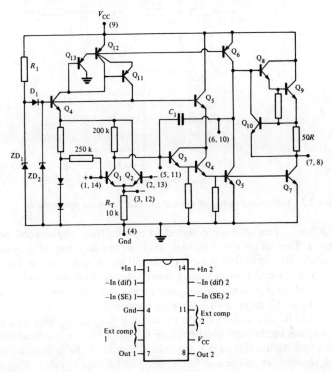

Figure 2.2   *Circuit and outline of the LM381/LM381A dual low-noise pre-amplifier*

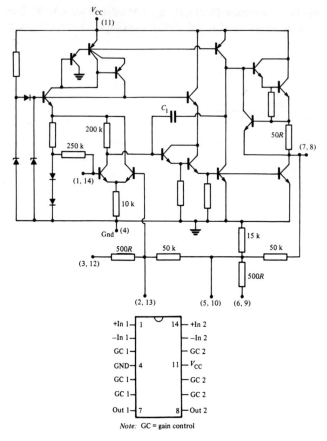

Figure 2.3   *Circuit and outline of the LM382 dual low-noise pre-amplifier*

All five of these ICs are designed to operate from single-ended power supplies. They all use the same basic amplifier circuitry, but differ in minor circuit details and in their pin-outs. They all incorporate internal feedback compensation and comprehensive power supply decoupler/regulator circuitry, and can give large output voltage swings and a wide power bandwidth. The various ICs differ in the following respects.

The LM381 and LM381A have provision for externally optimizing their noise figures and for adding external compensation (for narrow-band or low-gain applications). These ICs are normally used in the differential input configuration, but can be used in the 'single ended' input mode in ultra-low-noise applications.

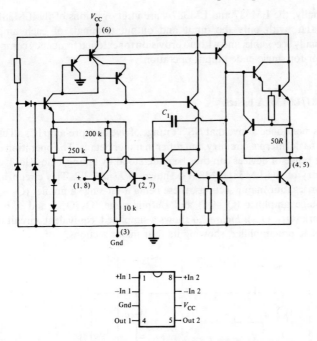

Figure 2.4   *Circuit and outline of the LM387/LM387A dual low-noise pre-amplifier*

	LM381	LM381A	LM382	LM387	LM387A
V supply	9 V–40 V	9 V–40 V	9 V–40 V	9 V–30 V	9 V–40 V
I quiescent (Typ)	10 mA	10 mA	10 mA	10 mA	10 mA
Power bandwidth (20 V pk-pk)	75 kHz	75 kHz	75 kHz	75 kHz	75 kHz
Supply rejection ratio at 1 kHz (typ)	120 dB	120 dB	120 dB	110 dB	110 dB
Equipment noise input figure, μV RMS Typ	0.5	0.5	0.8	0.8	0.65
Max.	1.0	0.7	1.2	1.2	0.9

Figure 2.5   *Performance characteristics of the five dual pre-amplifier ICs*

The LM382 has no provision for adding external compensation or for operation in the single-ended input configuration, but has a built-in resistor matrix that lets the user select a variety of closed-loop gain options and frequency-response characteristics.

Finally, the LM387 and LM387A are utility versions of the LM381 and LM381A, with only the input and output terminals of each amplifier externally accessible, and with no provision for external frequency compensation or for single-ended input operation.

LM381/LM381A basics

It was mentioned above that NSC's range of five dual pre-amp ICs all use the same basic internal circuitry, but differ in minor details. The operation of the entire range of devices can thus be understood by taking a close look at the circuitry of the LM381/LM381A shown in *Figure 2.2*. This circuit in fact comprises four major sections, these being a 1st-stage amplifier (Q_1–Q_2), a 2nd-stage amplifier (Q_3–Q_6) and output stage (Q_7–Q_{10}) and a biasing network (Q_{11}–Q_{15}). *Figure 2.6* shows a simplified 'equivalent' circuit of the complete pre-amplifier, showing its four major sections.

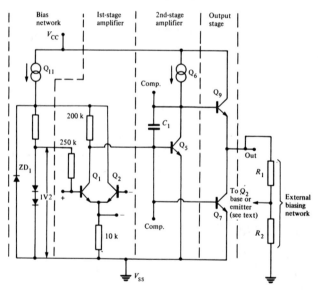

Figure 2.6   *Equivalent circuit of the LM381/LM381A amplifier*

The Q_1–Q_2 1st-stage input amplifier of the IC is powered via the internal biasing network, and has a biasing potential of 1.2 V permanently applied to Q_1 base via a 250 k series resistor. This 1st stage can be operated as either a differential or a single-ended amplifier (a differential stage generates 41% more noise than a single-ended stage).

When used in the differential mode, the Q_1–Q_2 amplifier must be balanced by feeding 1.2 V to Q_2 base via an external biasing network connected as shown. When used in the ultra-low-noise single-ended mode, Q_2 must be turned off by grounding its base, and Q_1 must be balanced by feeding 0.6 V to Q_2 emitter via the external biasing network. This 1st-stage amplifier gives a voltage gain of × 80 when used in the differential mode, or × 160 in the single-ended mode.

The 2nd-stage amplifier comprises common emitter stage Q_5 (with constant-current load Q_6), which is driven from the output of Q_1 via Darlington emitter follower Q_3–Q_4. This 2nd-stage amplifier gives an overall voltage gain of × 2000, and is internally compensated via C_1 to give unity gain at 15 MHz. This compensation provides stability at closed-loop gains of × 10 or greater. At lower gains an external capacitor can be wired in parallel with C_1 to provide suitable compensation.

The output stage of the amplifier comprises Darlington emitter follower Q_8–Q_9, which is provided with an active current sink via Q_7. Transistor Q_{10} provides short-circuit protection by limiting the output current to 12 mA.

The biasing network of the amplifier is designed to give a very high supply-signal rejection ratio (120 dB), and consists essentially of very high impedance constant-current generator Q_{11}–Q_{12}–Q_{13}, which is used to generate a ripple-free reference voltage across zener diode ZD_2. This reference voltage is then used to power the first two stages of the amplifier via Q_{14} and Q_{15}, and to provide internal biasing to Q_1 base.

Differential operation

The LM381 or LM381A IC can be operated in either the differential input or the single-ended input modes. Differential input operation is suitable for use in all general-purpose applications in which a 'good' low-noise performance is required. Single-ended input operation is recommended for use only in applications where an ultra-low-noise performance is needed.

To use a LM381 or LM381A pre-amp in the differential input mode the IC must first be biased so that its output takes up a positive quiescent value that is independent of variations in supply voltage, and this can be achieved by connecting potential divider R_1–R_2 between the output and the non-inverting input of the IC, as shown in *Figure 2.7*, thus forming a dc negative-feedback loop. The inverting input terminal of the IC (Q_1 base in *Figure 2.6*) is internally biased at about 1.2 V above zero; consequently, when R_1 and R_2 are connected as shown in *Figure 2.7*, dc negative feedback causes the non-inverting input terminal to take up a value equal to that of the inverting terminal (1.2 V). The amplifier output therefore attains a dc value of

$1V2 \times (R_1 + R_2)/R_2$, and can be set at any desired value by suitable choice of R_1/R_2 ratio. In practice, R_2 should have a value less than 250 k.

The *Figure 2.7* circuit can be made to act as a non-inverting ac-amplifier by simply ac-coupling the input signal to the non-inverting input terminal of the amplifier. In this configuration the circuit has an input impedance of about 250 k: input signals must be limited to 300 mV rms maximum, to avoid excessive distortion.

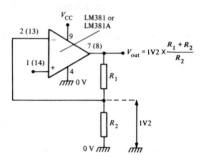

Figure 2.7   *Differential biasing of the LM381 or LM381A*

The dc voltage gain of the above circuit is determined by R_1 and R_2. If the desired ac-gain differs from the dc value, the desired ac-gain can be obtained by ac-shunting one or other of the bias network resistors. *Figure 2.8*, for example, shows the circuit of a low-noise $\times 100$ non-inverting amplifier. In this case the dc-gain is determined by R_1 and R_2 and is less than $\times 10$, but the ac-gain is determined mainly by R_1 and R_3, and approximates $\times 100$.

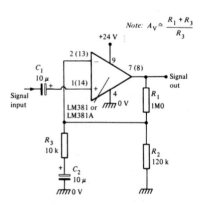

Figure 2.8   *Low-noise $\times 100$ non-inverting amplifier*

The above 'shunting' technique can easily be expanded to provide frequency-dependent ac-gain in various 'filter' applications. *Figure 2.9*, for example, shows the circuit of a low-noise phono pre-amp with RIAA equalization, and *Figure 2.10* shows a tape playback amplifier with National Association of Broadcasters (NAB) equalization.

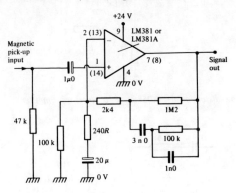

Figure 2.9   *Low-noise phono pre-amp (RIAA)*

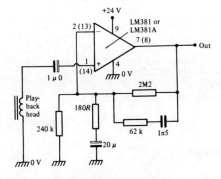

Figure 2.10   *Low-noise tape playback amplifier (NAB)*

The *Figure 2.7* circuit can be made to act as an inverting ac amplifier by ac grounding the non-inverting terminal and feeding the input signal to the inverting terminal via a gain-determining resistor, as shown in *Figure 2.11*. Here, bias resistors R_2 and R_3 give a dc-gain of about × 10, and thus set the quiescent output at + 12 V. The ac-gain, however, is determined by the R_3/R_1 ratio, and has a value of × 10 in this example: the input impedance roughly equals the value of R_1. Finally, *Figure 2.12* shows how the above circuit can be made to act as a unity-gain 4-input audio mixer by simply providing each of the four input channels with its own series-input resistor.

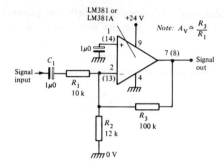

Figure 2.11   *Low-distortion (< 0.05%) × 10 inverting amplifier*

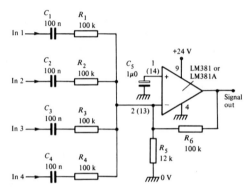

Figure 2.12   *Four-input unity-gain audio mixer*

Single-ended operation

The LM381A can be operated in the single-ended input mode in applications where an ultra-low-noise performance is needed. This mode can be understood with the aid of *Figure 2.13*, which shows (within the dotted lines) a simplified representation of the IC, together with external biasing components, etc.

In *Figure 2.13* the Q_1–Q_2 differential 1st-stage amplifier is shown powered via the internal 5V6 regulator, and has its Q_1 collector signal fed to the output via a dc-amplifier. The IC can be connected into the basic 'single-ended' configuration by simply grounding Q_2 base as shown, thereby disabling Q_2. Note in this case, however, that the circuit can no longer be dc biased via Q_2 base, so biasing must be achieved by using feedback to Q_2 emitter.

Suitable dc biasing can be obtained by connecting potential divider R_1–R_2 as shown, so that roughly 600 mV is developed across R_2 when the IC output is at the desired dc voltage level. Thus, if a quiescent output of $+12$ V is needed, R_1 and R_2 must give a dc voltage gain of $\times 20$. R_2 can, if desired, be shunted by R_3–C_1, to give an ac voltage gain that is greater than the dc value.

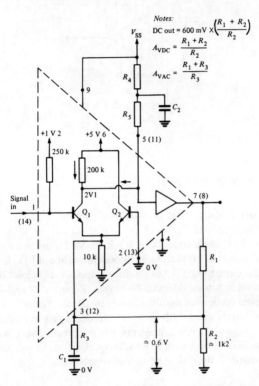

Notes:

$$\text{DC out} = 600 \text{ mV} \times \left(\frac{R_1 + R_2}{R_2}\right)$$

$$A_{VDC} = \frac{R_1 + R_2}{R_2}$$

$$A_{VAC} = \frac{R_1 + R_3}{R_3}$$

Figure 2.13 *LM381A with external components for single-ended varied current-density operation*

Note in the above biasing circuit that R_2 is in fact wired in parallel with the internal 10 k emitter resistor of Q_1 and thus causes the Q_1 emitter and collector 'current density levels' to increase above their normal values of about 15 μA. In practice, however, it can be shown that the noise generation of Q_1 varies with collector current density, and is minimum at a density of about 170 μA. Consequently, the circuit generates minimum noise when R_2 has a value of about 1k2. To prevent Q_1 collector from saturating at this current level the internal 200 k collector resistor of Q_1 must be bypassed and

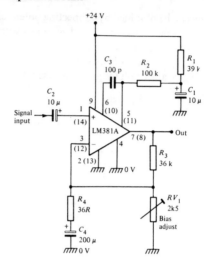

Figure 2.14   *Ultra-low-noise × 100 pre-amplifier*

the major part of the current provided via external load resistors R_4–R_5, which are decoupled via C_2.

The *Figure 2.13* 'single-ended' circuit is intended for use as a non-inverting amplifier only, and has a typical input impedance of about 10 k. Ideally, input signals to the circuit should have source impedances below 2k0, and all resistors should be low-noise metal-film types. *Figures 2.14* and *2.15* show a pair of practical versions of the ultra-low-noise circuit. *Figure 2.14* is a × 1000 amplifier and has the upper 3 dB point of its frequency curve limited to 10 kHz via C_3, and *Figure 2.15* is a magnetic-phono pre-amplifier with Record Industry Association of America (RIAA) equalization. In both cases, RV_1 is used to set the dc output voltage at half-supply value.

LM382 circuits

The internal circuitry of each half of the LM382 is identical to that of the LM381 except for the addition of a 5-resistor matrix and the elimination of certain terminal connections. The elimination of these terminals means that this IC cannot be used in the single-ended input mode and has no facility for external compensation, but the addition of the resistor matrix means that bias and filter network design can be greatly simplified; note that this matrix is specifically intended for use in applications in which the IC is powered from a 12 V supply.

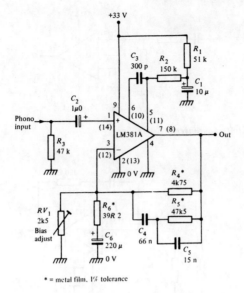

Figure 2.15 *Ultra-low-noise magnetic phono pre-amp with RIAA equalization*

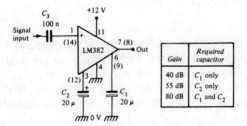

Figure 2.16 *LM382 used as a fixed-gain non-inverting amplifier with a 12 V power supply*

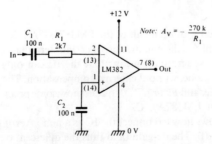

Figure 2.17 *A 40 dB inverting amplifier*

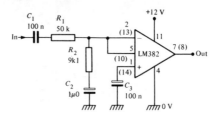

Figure 2.18   *Unity-gain inverting amplifier*

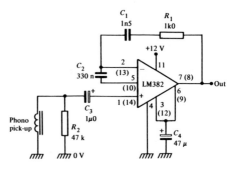

Figure 2.19   *Phono' pre-amp (RIAA)*

Figures 2.16 to *2.19* show various ways of using the LM382 with a 12 V supply. *Figure 2.16* shows how to use the IC as a non-inverting amplifier with an ac-gain of 40, 55 or 80 dB. *Figure 2.17* shows the circuit of an inverting amplifier with a gain of 40 dB, and *Figure 2.18* shows a unity-gain inverting amplifier. Finally, *Figure 2.19* shows a phono' pre-amplifier with RIAA equalization.

LM387 circuits

The internal circuitry of each half of the LM387/387A is identical to that of the LM381, except for the elimination of certain terminal connections. The elimination of these terminals means that the IC can only be used in the differential input mode, without external compensation. The IC is, neverthe-less, quite versatile, and *Figures 2.20* to *2.26* show some practical applications of the LM387 (or LM387A) IC.

Figure 2.20 shows how to connect the IC as a non-inverting amplifier with an ac-gain of 52 dB. The dc-gain (and thus the quiescent output voltage) is determined by R_1 and R_2, and the ac-gain is determined by R_1 and R_3. *Figure*

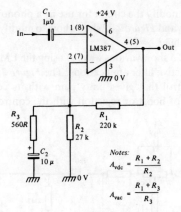

Figure 2.20 *LM387 non-inverting ac amplifier with a gain of 52 dB*

Notes:
$$A_{vdc} = \frac{R_1 + R_2}{R_2}$$

$$A_{vac} = \frac{R_1 + R_3}{R_3}$$

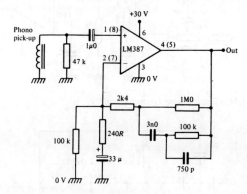

Figure 2.21 *LM387 phono pre-amp (RIAA)*

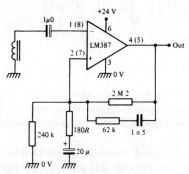

Figure 2.22 *LM387 tape playback amplifier (NAB)*

2.21 shows how to modify the circuit for use as a phono' pre-amplifier with RIAA equalization, and *Figure 2.22* shows how to modify it for use as a NAB tape playback amplifier.

Figures 2.23 to *2.26* show various ways of using the LM387 in the inverting amplifier mode in active filter applications. The *Figure 2.23* circuit is that of an active tone control that gives unity gain with its controls in the 'flat' position, or 20 dB of boost or rejection with the controls fully rotated.

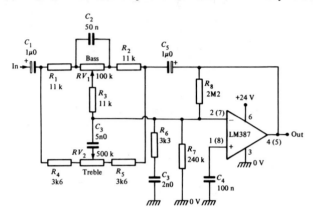

Figure 2.23   *LM387 active tone control circuit*

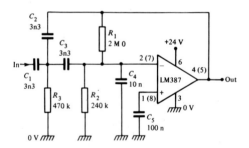

Figure 2.24   *'Rumble' filter*

The 'rumble' filter of *Figure 2.24* is actually a 2nd-order high-pass active filter that rejects signals below 50 Hz and does so with a slope of 12 dB/octave. The 'scratch' filter of *Figure 2.25* is a 2nd-order low-pass filter that rejects signals above 10 kHz.

Finally, the 'speech' filter of *Figure 2.26* consists of a 2nd-order high-pass and a 2nd-order low-pass filter wired in series, to give 12 dB/octave rejection to signals below 300 Hz or above 3 kHz.

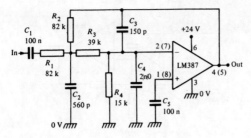

Figure 2.25   *'Scratch' filter*

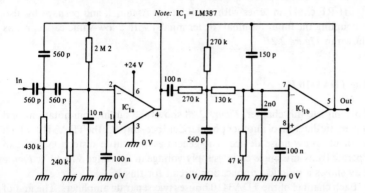

Figure 2.26   *'Speech (300 Hz to 3 kHz) filter*

LM381/382/387 usage hints

So far in this chapter we have looked at various circuits based on the
LM381/LM382/LM387 range of ICs. These ICs are high-gain, wide-band
devices, however, and some care must be taken in the construction of these
circuits if they are to work correctly. The two most frequently encountered
problems are those of radio frequency (RF)-instability and RF 'pickup'.

The RF-instability problem is usually caused by inadequate high-frequency power supply decoupling: note in *all* pre-amp circuits that the power
supply to the IC must be RF-decoupled by wiring a 100 n ceramic or 1n0
tantalum capacitor directly across the power supply pins of the IC.

The RF pickup problem manifests itself in the pickup and demodulation
of AM broadcast signals. This problem can usually be eliminated by wiring a

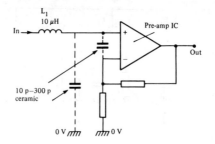

Figure 2.27   *'RF pick-up' elimination circuitry*

10 μH RF choke in series with the IC input terminal, and perhaps by also decoupling the input terminal (or terminals) with a low-value capacitor, as shown in *Figure 2.27*.

The TDA3410 IC

To complete this chapter, *Figure 2.28* shows the outline, pin notations and schematic diagram of another popular dual pre-amp IC, the TDA3410, which is an ultra-low-noise device that generates very little distortion and can operate from any single ended supply voltage in the range 8 to 30 V. *Figure 2.29* shows a practical application circuit for this IC.

Each channel of the TDA3410 houses two separate amplifiers. The first of these provides a fixed signal-voltage gain of 30 dB ($\times$ 32), and at any given

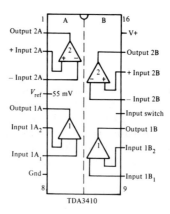

Figure 2.28   *Outline, pin notations and schematic of the TDA3410 ultra-low-noise dual pre-amp IC*

moment can accept one or other of two input signals, which are applied to pins 7(9) and 6(10); the desired input signal can be selected by applying a suitable control voltage to input switch pin-12 of the IC. When this pin is grounded (as in *Figure 2.29*), internal switches clamp pin 6(10) to ground and pin 7(9) is enabled; if pin-12 is taken above 4 V, the switching action is reversed, and pin 7(9) is grounded and pin 6(10) is enabled.

The two input terminals of the first amplifier each have an input impedance of about 80 k. This amplifier stage can provide maximum undistorted output signal swings of about 2 V pk-to-pk, and should thus not be driven by input voltages greater than 63 mV pk-to-pk. The pin 5(11) output of the first amplifier can be wired directly to the pin 2(14) input of the second amplifier, if desired, as shown in *Figure 2.29*.

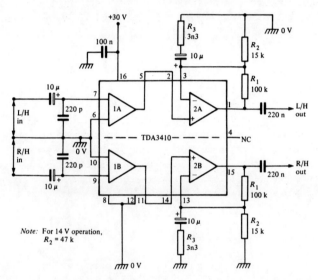

Figure 2.29   *TDA3410 stereo pre-amp circuit with an overall voltage gain of 60 dB* (× 1000)

The second amplifier can be used like a normal op-amp, as shown in *Figure 2.29*, where its dc voltage gain (bias) is set via R_1 and R_2, and the ac-gain is set via R_1 and R_3. In the example shown, the ac-gain is set at about 30 dB, and under this condition the amplifier's power bandwidth exceeds 22 kHz. This amplifier can provide maximum undistorted (THD less than 0.05%) output signals of about 6 V pk-to-pk.

Thus, in the application circuit of *Figure 2.29*, the IC gives an overall voltage gain of about 60 dB (× 1000), and thus gives a 1 V pk-to-pk output from a 1 mV pk-to-pk input. Under this condition the circuit's 0.5 dB frequency response extends from 25 Hz to 20 kHz, the signal distortion (THD) is less than 0.05%, and the signal-to-noise ratio is better than 65 dB; the equivalent total input noise of each channel of the amplifier is less than 0.5 mV when using an input source resistance less than 1k0.

When considering use of the TDA3410, note that this IC incorporates a voltage regulator (to give a very high 'supply voltage rejection' figure) that also provides a 55 mV voltage reference that is brought out at pin-4. This reference voltage can be used to bias the second amplifier by connecting a 22 k resistor between pins 2(14) and 4, if desired.

3 Audio power amplifier circuits

An 'ideal' audio power amplifier can be simply defined as a circuit that can deliver audio power into an external load without generating significant signal distortion and without overheating or consuming excessive quiescent current. Circuits that come very close to this ideal can easily be built, using modern integrated circuits.

Simple audio power amplifiers with outputs up to only a few hundred milliwatts can be easily and cheaply built using little more than a standard op-amp and a couple of general-purpose transistors. For higher power levels, a wide range of special-purpose 'single' or 'dual' audio power amplifier ICs are readily available, and can provide maximum outputs ranging from a few hundred milliwatts to roughly 20 W. The specific IC chosen for a given application depends mainly on the constraints of the available power supply voltage and on the required output power level or levels. A wide selection of practical IC-based circuits with maximum power-output ratings up to about 5 W are described in this chapter; circuits with power ratings in the range 5 to 22 W are dealt with in Chapter 4.

Low-power op-amp circuits

The popular 741 general-purpose operational amplifier can supply peak output currents of at least 10 mA, and can provide peak output voltage swings of at least 10 V into a 1k0 load when powered from a dual 15 V supply. This IC can thus supply peaks of about 100 mW into a 1k0 load under this condition, and can easily be used as a simple low-power audio amplifier, as shown in *Figures 3.1* and *3.2*.

Figure 3.1 shows how to use the 741 op-amp as a low-power amplifier in conjunction with a dual power supply. The external load is direct-coupled

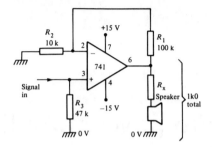

Figure 3.1   *Low-power amplifier using dual power supplies*

between the op-amp output and ground, and the two input terminals are ground-referenced. The op-amp is used in the non-inverting mode, and has a voltage gain of × 10 (R_1/R_2) and input impedance of 47 k (R_3).

Figure 3.2 shows how to use the circuit with a single-ended power supply. In this case the external load is ac-coupled between the output and ground, and the output is biased to a quiescent value of half-supply volts (to give maximum output voltage swing) via the R_1–R_2 potential divider. The op-amp is operated in the unity-gain non-inverting mode, and has an input impedance of 47 k (R_3).

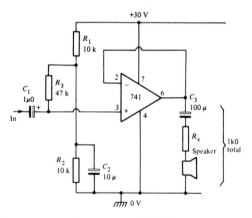

Figure 3.2   *Low-power amplifier using a single-ended power supply*

Note in the above two circuits that the external load must have an impedance of at least 1k0. If the external 'speaker has an impedance lower than this value, resistor R_x can be connected as shown to raise the impedance to the 1k0 value. R_x inevitably reduces the amount of power reaching the actual 'speaker'.

Boosted-output op-amp circuits

The available output current (and thus power) of an op-amp can easily be boosted by wiring a complementary emitter follower between its output and its non-inverting input terminal, as shown in *Figure 3.3*. Note that this circuit is configured to give an overall voltage gain of unity, but that the Q_1 and Q_2 base-emitter junctions are both wired into the circuit's negative feedback loop, so that their effective forward voltage values (about 600 mV) are reduced by a factor equal to the open-loop voltage gain of the op-amp. Thus,

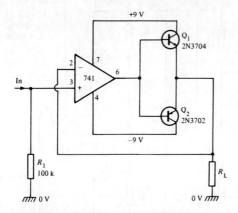

Figure 3.3 *Basic boosted output current unity-voltage-gain op-amp circuit*

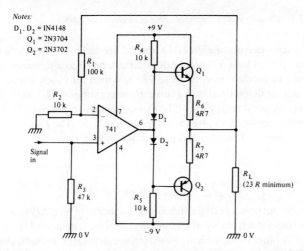

Figure 3.4 *Op-amp power amplifier using dual supplies (power out = 280 mW max)*

if this gain is $\times 10{,}000$ the effective forward voltages of Q_1 and Q_2 are each reduced to a mere $6\,\mu V$, and the circuit generates negligible signal distortion.

In practice, op-amp open-loop voltage gain falls off at a rate of about 20 dB/octave, so although the signal distortion of the *Figure 3.3* amplifier may be insignificant at 10 Hz, it can rise to objectionable levels at, say, 10 kHz. This problem can be overcome by applying a slight forward bias to Q_1 and Q_2 as shown in *Figures 3.4* and *3.5*, so that their forward voltage values are reduced to near-zero and distortion is minimized.

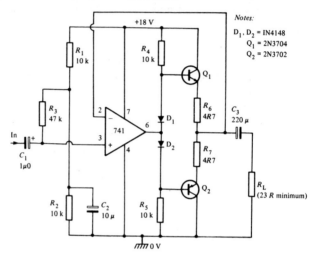

Figure 3.5   *Op-amp power amplifier using a single-ended supply*

The specific circuits of *Figures 3.4* and *3.5* are designed to produce output currents up to at least 350 mA peak or 50 mA rms into a minimum load of 23 Ω, i.e., to produce powers up to 280 mW rms into such a load. These limitations are determined by the current/power ratings of Q_1 and Q_2, and by the power supply voltage values. The *Figure 3.4* circuit is designed for use with dual power supplies, and gives a voltage gain of $\times 10$. The *Figure 3.5* circuit uses a single-ended supply, and gives unity voltage gain.

IC power amplifier basics

If audio output powers in the approximate range 200 mW to 20 W are needed, the most cost-effective way of getting them is to use a dedicated IC to do the job. A wide range of such ICs are available, in either single or dual form. Most of these ICs take the effective form of an op-amp with a complementary

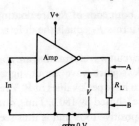

Figure 3.6   *An amplifier connected in the single-ended output mode gives a peak output of* V^2/R *W*

emitter-follower output stage (like *Figures 3.4* and *3.5*); they have differential input terminals and can provide high output current/power, but consume a low quiescent current.

When an IC power amplifier is connected in the single-ended output mode, as shown in *Figure 3.6*, the peak available output power equals V^2/R, where V is the peak available output voltage. Note, however, that available output power can be increased by a factor of four by connecting a pair of amplifier ICs in the 'bridge' configuration shown in *Figure 3.7*, in which the peak available load power equals $2V^2/R$. This power increase can be explained as follows.

In the single-ended amplifier circuit of *Figure 3.6* one end of R_L is grounded, so the peak voltage across R_L equals the voltage value on point A. In *Figure*

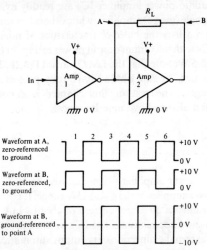

Figure 3.7   *A pair of amplifiers connected in the bridge mode give a peak output of* $2V^2/R$ *Watts, i.e., four times the power of a single-ended circuit*

3.7, on the other hand, both ends of R_L are floating and are driven in anti-phase, and the voltage across R_L equals the difference between the A and B values.

Figure 3.7 shows the circuit waveforms that are applied to the load when fed with a 10 V pk-to-pk square-wave input signal. Note that although waveforms A and B each have peak values of 10 V relative to ground, the two signals are in anti-phase (shifted by 180°). Thus, during period 1 of the drive signal, point B is 10 V positive to A and is thus seen as being at + 10 V. In period 2, however, point B is 10 V negative to point A, and is thus seen as being at − 10 V. Consequently, if point A is regarded as a zero voltage reference point, it can be seen that the point B voltage varies from + 10 V to − 10 V between periods 1 and 2, giving a total voltage change of 20 V across R_L. Similar changes occur in subsequent waveform periods.

Thus, the load in a 10 V bridge-driven circuit sees a total voltage of 20 V peak-to-peak, or twice the single-ended input voltage value, as indicated in the diagram. Since doubling the drive voltage results in a doubling of drive current, and power is equal to the *V–I* product, the bridge-driven circuit thus produces four times more power output than a single-ended circuit. We will look at a range of IC-based single-ended and bridge-driven power amplifier circuits throughout the rest of this and the following chapter.

Practical ICs

A large range of audio power amplifier ICs are readily available. Some of these ICs house a single (mono) amplifier, while others house a pair (a dual) of amplifiers. *Figure 3.8* lists the basic characteristics of nine popular audio power amplifier ICs with maximum output power ratings in the approximate range 325 mW to 5.5 W; note that the LM831 and TDA2822 are dual types, and that only the LM380 and LM384 have fully protected (short-circuit proof) output stages. The rest of this chapter is devoted to detailed descriptions of each of the above nine IC types.

LM386 basics

The LM386 audio power amplifier (manufactured by NSC) is designed for operation with power supplies in the 4 V to 15 V range. It is housed in an 8-pin DIL package, consumes a quiescent current of about 4 mA, and is ideal for use in battery-powered applications. The IC's voltage gain is variable from × 20 to × 200 via external connections, its output automatically centres on a quiescent half-supply voltage value, and it can feed several hundred milliwatts into an 8 Ω load when operated from a 12 V supply. Its differential input

Device number	Amplifier type	Maximum output power	Supply voltage range	Distortion, into 8R0	Input impedance	Voltage gain	Bandwidth	Quiescent current
LM 386	Mono	325 mW into 8R0	4 to 15 V	0.2%, $V_S = 6$ V $P_O = 125$ mW	50 k	26 dB	300 kHz	4 mA
LM 389	Mono, and 3-transistor array	325 mW into 8R0	4 to 15 V	0.2%, $V_S = 6$ V $P_O = 125$ mW	50 k	26 dB	250 kHz	6 mA
LM 831	Dual	220 mW/channel into 4R0	1.8 to 6 V	0.25%, $V_S = 3$ V $P_O = 50$ mW	25 k	46 dB	20 Hz–20 kHz	6 mA
LM 390	Mono	1 W into 4R0	4 to 9 V	0.2%, $V_S = 6$ V $P_O = 0.5$ W	50 k	26 dB	300 kHz	10 mA
LM 388	Mono	1.5 W into 8R0	4 to 12 V	0.1%, $V_S = 12$ V $P_O = 0.5$ W	50 k	26 dB	300 kHz	16 mA
TDA 2822	Dual	1 W/channel into 8R0	1.8 to 15 V	0.3%, $V_S = 9$ V $P_O = 0.5$ W	100 k	40 dB	120 kHz	6 mA
TBA 820 M	Mono	2 W into 8R0	3 to 16 V	0.4%, $V_S = 9$ V $P_O = 0.5$ W	5M0	34 dB	20 Hz–20 kHz	4 mA
LM 380	Mono, with protected output	3 W into 4R0	8 to 22 V	0.2%, $V_S = 18$ V $P_O = 2$ W	150 k	34 dB	100 kHz	7 mA
LM 384	Mono, with protected output	5.5 W into 8R0	12 to 26 V	0.25%, $V_S = 22$ V $P_O = 4$ W	150 k	34 dB	450 kHz	8.5 mA

Figure 3.8 *Basic details of the nine ICs described in this chapter*

terminals are both ground-referenced, and have typical input impedances of 50 k.

Figure 3.9 shows the LM386's internal circuit. Here, Q_1 to Q_6 form a differential amplifier in which both inputs are tied to ground via 50 k resistors (R_1 and R_2) and the output (from Q_3) is direct-coupled to the input of common emitter amplifier Q_7. The Q_7 collector signal is direct-coupled to the IC's output terminal via class-B unity-gain power amplifier stage Q_8–Q_9–Q_{10} which, to minimize internal volt-drops and maximize the available output power, is not provided with overloaded protection circuitry.

LM386 applications

The LM386 is very easy to use. Its voltage gain equals double the pin-1 to pin-5 impedance value (15 k in *Figure 3.9*) divided by the impedance between the emitters of Q_1 and Q_3 ($R_5 + R_6$). Thus, the IC can be used as a minimum-parts amplifier with an overall voltage gain of $\times 20$ (2×15 k/1.5 k) by using the simple connections shown in *Figure 3.10*, where the load is ac-coupled to the IC output via C_2, and the input signal is fed to the non-inverting terminal via RV_1. Note that C_1 is used to RF-decouple the +ve supply pin (pin-6), and

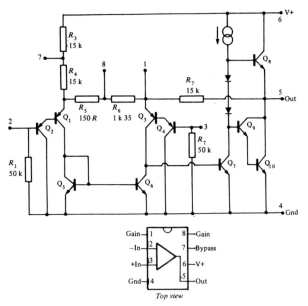

Figure 3.9   *Internal circuit and pin connections of the LM386 low-voltage audio power amplifier*

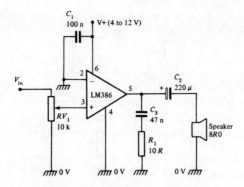

Figure 3.10   *Minimum-parts LM386 amplifier with* $A_v = 20$

R_1–C_3 is an optional Zobel network that ensures high frequency (HF) stability when feeding an inductive speaker load.

Figure 3.11 shows how the above circuit can be modified to give an overall voltage gain of × 200, by using C_4 (between pins 1 and 8) to effectively short-circuit the internal 1k35 resistor of the IC. Alternatively, *Figure 3.12* shows how the gain can be set at × 50 by wiring a 1k2 resistor (R_2) in series with C_4.

The voltage gain of the LM386 can also be varied by shunting the effective value of the internal 15 k pin-5 to pin-1 feedback resistor. *Figure 3.13* shows how to shunt this resistor with C_4–R_2, to give 6 dB of bass boost at 85 Hz, to compensate for the poor bass response of a cheap speaker.

Finally, *Figure 3.14* shows how the LM386 amplifier can be modified for use as a built-in amplifier in an AM radio. Here, the detected AM signal is fed to the non-inverting input of the IC via volume control RV_1, and is RF-

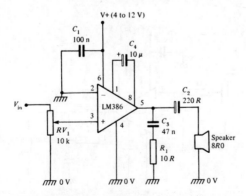

Figure 3.11   *LM386 amplifier with* $A_v = 200$

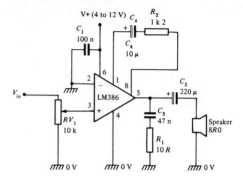

Figure 3.12   *LM386 amplifier with A$_v$ = 50*

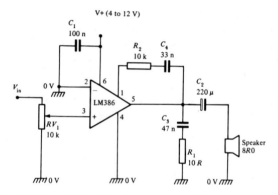

Figure 3.13   *LM386 amplifier with 6 dB of bass-boost at 85 Hz*

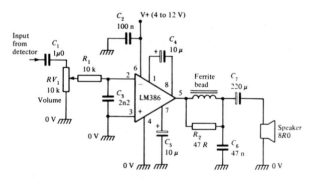

Figure 3.14   *AM-radio power amplifier*

decoupled via R_1–C_3; any residual RF signals are blocked from the load via a ferrite bead. The voltage gain of the amplifier is set at × 200 via C_1. Note that this circuit is provided with additional power-supply ripple rejection by wiring C_5 between pin-7 and ground; this ripple-rejection capacitor can also be used with the *Figure 3.10* to *3.13* circuits if required.

LM389 circuits

The LM389 (*Figure 3.15*) contains an array of three independently accessible wide-band npn transistors on the same substrate as an audio power amplifier that is almost identical to that of the LM386. The three npn transistors have closely matched characteristics, can be operated with collector currents in the range 1 μA to 25 mA, at frequencies up to 100 MHz, and each have typical current gain values of × 275. The IC can use any power supply in the range 4 to 15 V.

Figure 3.16 shows the IC's basic connections. The internal power amplifier is used in the same way as the LM386, with its voltage gain controlled by C_4 and R_x, between pins 4 and 12. If C_4 and R_x are absent, the power amplifier

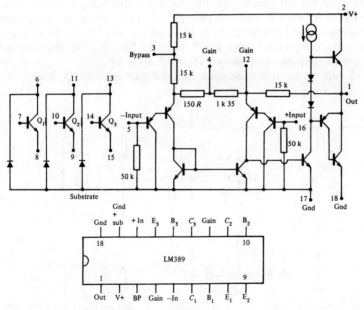

Figure 3.15 *Circuit and outline of the LM389 low-voltage audio power amplifier with npn transistor array*

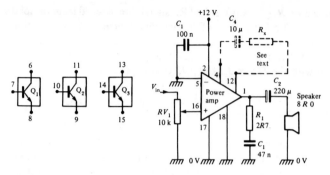

Figure 3.16   *Basic circuit connections of the LM389 IC*

voltage gain is × 20; if they are fitted and R_x has a value of 1k2, the gain is × 50; if R_x is a short circuit, the gain rises to × 200. The power amplifier can be used as either an inverting or non-inverting unit by connecting the external signal to the appropriate input terminal. Note that Q_1, Q_2 and Q_3 of the IC are independently accessible.

Figures 3.17 and *3.18* show practical applications of the LM389, making use of the internal transistors. In the phono' amplifier of *Figure 3.17*, which is intended for use with a ceramic pickup, Q_3 acts as a voltage following input buffer giving an input impedance of about 800 k, and Q_1–Q_2 are used to make an active tone control network with its output feeding to the non-inverting input of the power amplifier via volume control RV_3.

Finally, in the white noise generator circuit of *Figure 3.18*, Q_3 is wired as a

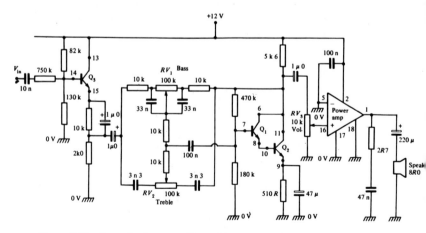

Figure 3.17   *Ceramic phono amplifier, with tone controls, using an LM389*

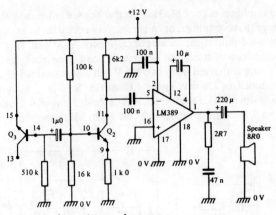

Figure 3.18 *LM389 white-noise sound generator*

noise-generating zener diode, and this noise signal is amplified via Q_2 and then fed to the inverting input terminal of the power amplifier, which is wired in the × 200 voltage-gain mode.

LM831 circuits

Then LM831 is a dual power-amplifier IC that is specifically designed for very low voltage operation; it can use supplies in the 1.8 to 6 V range. Its two independent amplifiers give good low-noise and low-distortion performances, and generate minimal RF radiation, thus enabling the IC to be used in close proximity to an AM receiver. The IC is housed in a 16-pin DIL package, using the pin notations shown in *Figure 3.19*.

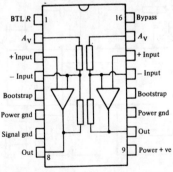

Figure 3.19 *Outline, basic circuit and pin notations of the LM831 dual low-voltage audio power amplifier*

The two amplifiers of the LM831 can either be used independently to make a low-voltage stereo amplifier, or can be interconnected in the bridge mode to make a boosted-output mono amplifier. *Figures 3.20* and *3.21* show the circuit connections of these two options. When these circuits are powered from a 3 V supply derived from two 1.5 V cells, each channel of the stereo amplifier can deliver 220 mW into a 4 Ω speaker load (and will give a 3 dB signal bandwidth of 50 Hz to 20 kHz), and the bridge amplifier can deliver 440 mW into an 8 Ω load (and gives a 20 Hz to 20 kHz bandwidth).

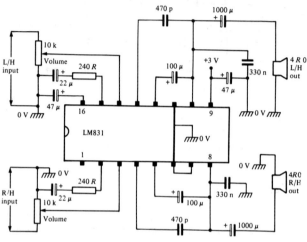

Figure 3.20   *LM831 stereo amplifier*

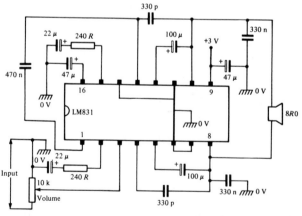

Figure 3.21   *Bridge-connected LM831 amplifier*

When constructing these two circuits note that, in the interest of adequate circuit stability, the PCB must be laid out with large earth planes, and the pin-9 decoupling capacitor must be as close to the IC as possible and must have a minimum value of 47 μF; the two 0.33 μF decoupling capacitors must also be as close as possible to the IC.

LM390 circuits

The LM390 is described in manufacturer's literature as a 1-watt battery-operated audio power amplifier. It is designed for operation from 4 V to 9 V power supplies, and can feed 1 W into a 4R0 load when using a 6 V supply. *Figure 3.22* shows the internal circuit and pin connections of this IC, which is internally very similar to the LM386 described earlier, but has its output stage modified to give the maximum possible output voltage swing. The device is housed in a 14-pin DIL package with an internal heat sink connected to pins 3–4–5 and 10–11–12.

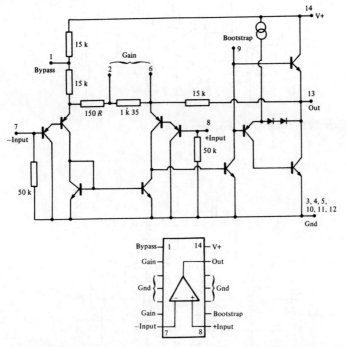

Figure 3.22 *Internal circuit and pin connections of the LM390 1 W battery-operated audio power amplifier IC*

The IC's overall voltage gain is internally set at × 20, but can be increased to × 200 by wiring a shunt capacitor between pins 2 and 6. The IC inputs are ground referenced, and the output automatically self-biases to a quiescent value of half-supply volts when the output stage is suitably dc-biased via external resistors wired between pins 9 and 14. *Figures 3.23* to *3.27* show some practical LM390 applications.

Figure 3.23 shows one way of using the IC as a 1 W amplifier driving a 4R0 load from a 6 V supply. R_1 and R_2 are wired in series between the positive supply line and pin-9, to give dc biasing to the IC's output stage. Note that the R_1–R_2 junction is bootstrapped from the IC's output via C_2, to raise the ac-impedance of R_2 to a value far greater than its dc value. The overall voltage gain of the LM390 is internally determined in the same way as in the LM386,

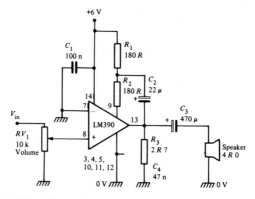

Figure 3.23   *LM390 1 W amplifier with* $A_v = 20$

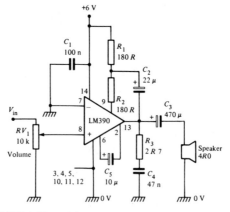

Figure 3.24   *LM390 1 W amplifier with* $A_v = 200$

and thus equals × 20 in *Figure 3.23*. *Figure 3.24* shows how the gain can be increased to × 200 by simply wiring C_5 between pins 2 and 6.

Figure 3.25 shows an alternative way of using the LM390. Here, DC current is fed to pin-9 of the IC via the speaker and R_1; note that R_1 is bootstrapped via C_2, so this circuit gives a performance similar to that of *Figure 3.23*, but does so with a saving of two components.

Figure 3.26 shows how to connect a pair of LM390 ICs in the bridge configuration, to give 2.5 W of drive to a direct-coupled 4R0 load when using

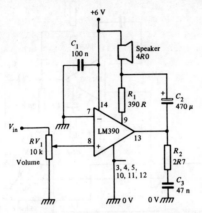

Figure 3.25   *LM390 1 W amplifier with* $A_v = 20$ *and load returned to* + *ve supply*

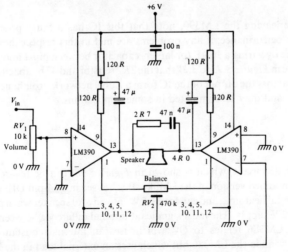

Figure 3.26   *LM390 bridge amplifier delivers 2.5 W into a 4R0 load*

a 6 V supply. Pre-set pot RV_2 is used to balance the quiescent outputs of the two ICs and thus minimize the circuit's quiescent current consumption.

Finally, *Figure 3.27* shows how to use a single LM390 IC to make a simple 2-way intercom circuit. Note that C_5–R_4 are used to provide the IC with an overall voltage gain of × 300 (15 k/51 R).

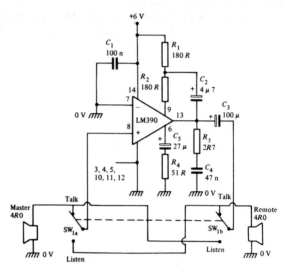

Figure 3.27   *LM390-based intercom*

Before leaving the LM390, note that this IC has a fairly poor ripple-rejection performance; if any problems are met in this respect they can be overcome by wiring a 10 μF (or larger) capacitor between pin-1 and ground. Also note in *Figures 3.23* to *3.27* that the 2R7 resistor and 47 n capacitor wired in series across the output of the IC form a Zobel network, to enhance circuit stability, and may be eliminated in some applications.

LM388 circuits

The LM388 circuit, which is shown in *Figure 3.28*, can be regarded as a slightly modified version of the LM386. It is housed in a 14-pin DIL package with internal heat sink, and can feed 1.5 W into an 8R0 speaker when powered from a 12 V supply. The most significant internal difference between this IC and the LM386 relates to Q_7, which uses an internal constant-current collector load in the LM386 but which uses an external load in the LM388. This external load feature greatly increases the versatility of the IC.

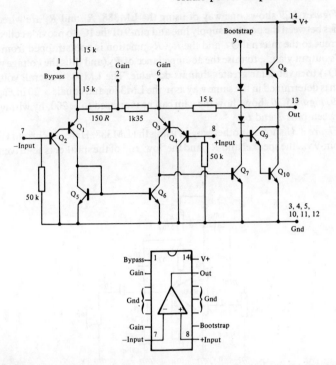

Figure 3.28   *Internal circuit and pin connections of the LM388 1.5 W audio power amplifier*

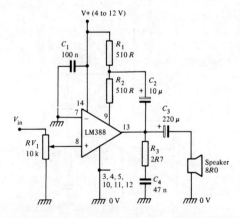

Figure 3.29   *LM388 with a gain of × 20 and load returned to ground*

Figures 3.29 shows one way of using the LM388. R_1 and R_2 are wired in series between the positive supply line and pin-9 of the IC, to provide collector current to the internal Q_7, and the R_1–R_2 junction is bootstrapped from the IC's output via C_2, to raise the ac-impedance of R_2 (and thus the voltage gain of Q_7) to a value far greater than its dc value. The LM388's overall voltage gain is determined in the same way as in the LM386, and equals × 20 in *Figure 3.29*. *Figure 3.30* shows how the gain can be increased to × 200, by wiring C_5 between pins 2 and 6.

Figure 3.31 shows another way of using the LM388. Here, dc current is fed to pin-9 via the speaker and R_1, and the 'low' end of the speaker is ac-driven by

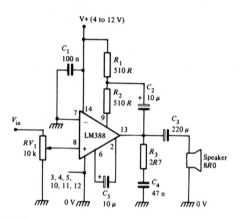

Figure 3.30   *LM388 with a gain of × 200 and load returned to ground*

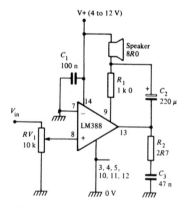

Figure 3.31   *LM388 with a gain of × 20 and load returned to + ve supply*

the amplifier's output, thus bootstrapping R_1 and giving it a high ac-impedance value. This circuit thus gives a performance similar to that of *Figure 3.29*, but does so with a saving of two components.

Finally, *Figure 3.32* shows how to connect a pair of LM388 ICs in the bridge configuration, to provide 4 W of drive to a direct-coupled 8R0 speaker load when using a 12 V power supply. Pre-set pot RV_2 is used here to set the quiescent output of the two ICs at identical values, to minimize the circuit's quiescent current consumption.

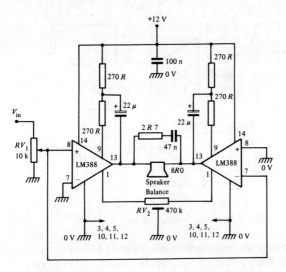

Figure 3.32 *LM388 bridge amplifier delivering 4 W to an 8R0 load*

Before leaving the LM388, note that (like the LM390) this IC has a fairly poor supply line ripple rejection performance, and if any problems are met in this respect they can be overcome by wiring a 10 μF (or larger) capacitor between pin-1 and ground.

TDA2822 circuits

The TDA2822 is a versatile dual amplifier that can use any dc supply in the 1.8 V to 15 V range; it can be powered from a 3 V supply and used to drive headphones at 20 mW per 32 Ω channel, or from a 9V supply and used to drive 8 Ω speakers at 1 W per channel.

The TDA2822 is housed in an 8-pin DIL package (see *Figure 3.33*), and uses

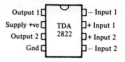

Figure 3.33   *Outline and pin notations of the TDA2822 dual amplifier IC*

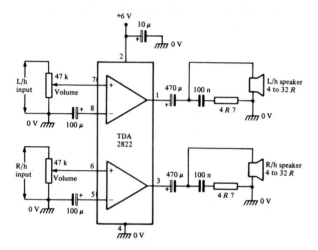

Figure 3.34   *TDA2822 stereo amplifier circuit*

the minimum of external components. *Figure 3.34* shows how it can be used as a stereo speaker or headphone amplifier circuit that is powered from a 6 V supply.

TBA820M circuits

The manufacturers describe this device as a low-power amplifier that is capable of generating a few hundred milliwatts in a $4R0$ to $16R$ speaker load, although it can, in fact, generate as much as 2 W in an $8R0$ load. The IC is housed in an 8-pin DIL package, can operate from supplies as low as 3 V, and features low quiescent current, good ripple rejection, and low crossover distortion.

Figure 3.35 shows the outline and pin notations of the TBA820M IC, plus a practical application circuit for the device. Here, R_2 determines the voltage gain of the IC, and R_3–C_6 from a Zobel network across the load speaker. This circuit can use a maximum supply voltage of 16 V with a $16R$ speaker, 12 V with an $8R0$ speaker, or 9 V with a $4R0$ speaker.

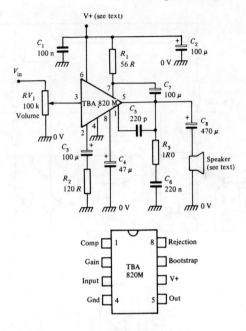

Figure 3.35 *TBA820M low-power audio amplifier circuit*

LM380/LM384 circuits

The LM380 circuit (*Figure 3.36*) is probably the best known of all power amplifier ICs. It can work with any supply voltage in the range 8 V to 22 V, and can deliver 2 W into an 8R0 load when operated with an 18 V supply but needs a good external heat sink to cope with this power level. Its differential input terminals are both ground referenced, and the output automatically sets at a quiescent value of half-supply volts. Its voltage gain is fixed at × 50 (34 dB), the output is short-circuit proof, and the IC is provided with internal thermal limiting.

The LM384 is simply an up rated version of the LM380, capable of operating at supply values up to 26 V and of delivering 5.5 W into an external load. Both types of IC are housed in a 14-pin DIL package, in which pins 3–4–5 and 10–11–12 are intended to be thermally coupled to an external heat sink.

To conclude this chapter, *Figures 3.37* to *3.40* show some practical applications of these two audio power amplifier ICs. *Figure 3.37* shows how to use either IC as a simple (× 50) amplifier with enhanced (via C_2) ripple

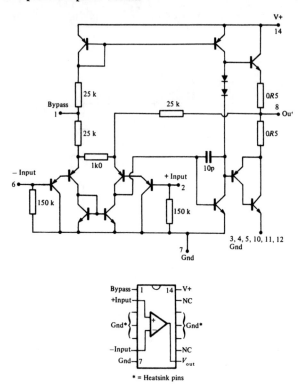

Figure 3.36   *Circuit and pin connections of the LM380 2 W and LM384 5 W audio power amplifier ICs*

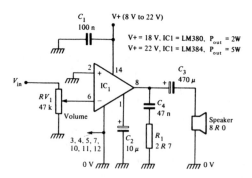

Figure 3.37   *2 W or 5 W amplifier with simple volume control and ripple rejection*

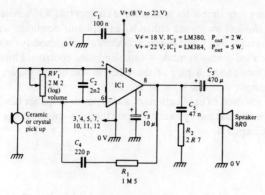

Figure 3.38 *2 W or 5 W phono amplifier with RIAA equalization*

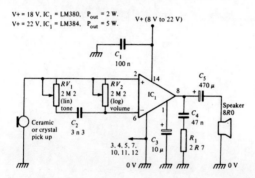

Figure 3.39 *2 W or 5 W phono amplifier with common mode volume and tone controls*

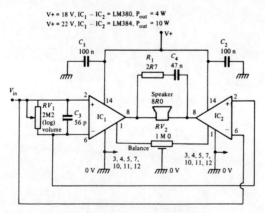

Figure 3.40 *4 W or 10 W bridge-configured amplifier*

rejection and a very simple form of volume control (via RV_1). Alternatively, *Figure 3.38* shows how to use either IC as a phono' amplifier with RIAA equalization (via $R_1–C_4$), and *Figure 3.39* shows how to modify the circuit for use with so-called common mode volume and tone controls. Finally, *Figure 3.40* shows how to use a pair of these ICs in the bridge mode, to give a maximum output of either 4 or 10 W.

A further selection of applications of audio power amplifier ICs (including dual types), with maximum power ratings in the approximate range 5 to 20 W, is given in Chapter 4.

4 High-power audio amplifiers

In Chapter 3 we looked at a selection of popular audio power amplifier ICs with maximum output power ratings in the approximate range 325 mW to 5.5 W and showed a variety of ways of using these devices in practical circuit applications. In this chapter we continue the audio power amplifier theme by looking at a further selection of these ICs and their application circuits, but in this case we deal with devices with maximum output power ratings in the range 5 W to 25 W.

Figure 4.1 gives basic details of the IC types that are dealt with in this chapter. Note that five of these ICs (the LM377, LM378, LM379, LM2879, and TDA2004) are dual types which each house a pair of independently accessible amplifiers, but that the TDA2005M is a bridge type which houses a pair of amplifiers that are permanently wired in the bridge or power boosting configuration.

Throughout this chapter all ICs are dealt with in the order in which they are listed in *Figure 4.1*. Practical application circuits are given for each IC type, but in some cases only very brief descriptions are given of individual IC circuit theory.

LM377/378/379 circuits

NSC produce a range of dual audio power amplifier ICs for use in stereo amplifiers and in bridge-configured mono amplifier applications. The best known of these devices are the LM377 dual 2 W, the LM378 dual 4 W, and the LM379 dual 6 W amplifiers. *Figures 4.2* and *4.3* show the outlines and pin notations of these devices, and *Figure 4.4* shows the approximate performance characteristics of the three ICs.

The LM377/378/379 range of ICs all have similar internal circuits, with

Device number	Amplifier type	Maximum output power	Supply voltage range	Distortion into 8 R 0	Input impedance	Voltage gain	Bandwidth	Quiescent current
LM 377	Dual	2.5 W/channel	10 to 26 V	0.1% at 2 W/channel	3 MΩ	34 dB	50 kHz	15 mA
LM 378	Dual	4 W/channel	10 to 35 V	0.1% at 2 W/channel	3 MΩ	34 dB	50 kHz	15 mA
LM 379	Dual	6 W/channel	10 to 35 V	0.2% at 4 W/channel	3 MΩ	34 dB	50 kHz	15 mA
TBA 810S TBA 810P	Mono	6 W into 4R0'	4 to 20 V	0.3% at 2.5 W	5 MΩ	37 dB	40 Hz – 15 kHz 40 Hz – 20 kHz	12 mA
LM 383 TDA 2003	Mono	7 W into 4R0	5 to 25 V	0.2% at 4 W	150 k	40 dB	30 kHz	45 mA
LM2002 TDA 2002	Mono	8 W into 2R0	5 to 20 V	0.1% at 4 W	150 k	40 dB	100 kHz	45 mA
LM 2879	Dual	9 W/channel	10 to 35 V	0.04% at 4 W/channel	3 MΩ	34 dB	50 kHz	12 mA
TDA 2004	Dual	9 W/channel	8 to 18 V	0.2% at 4 W/channel	200 k	50 dB	35 kHz	65 mA
TDA 2006	Mono	12 W into 4R0	±6 to ±15 V	0.1% at 4 W	5 MΩ	30 dB	150 kHz	40 mA
TDA 2030	Mono	18 W into 4R0	±6 to ±18 V	0.1% at 8 W	5 MΩ	30 dB	140 kHz	40 mA
TDA 2005M	Bridge	20 W into 4R0	6 to 18 V	0.25% at 12 W	100 k	50 dB	20 kHz	75 mA
TDA 1520	Mono	22 W into 4R0	15 to 40 V	0.01% at 16 W	20 k	30 dB	20 kHz	54 mA
LM 1875	Mono	25 W into 4R0	20 to 60 V	0.015% at 20 W	1 MΩ	26 dB	70 kHz	70 mA

Figure 4.1   *Basic details of the ICs described in this chapter*

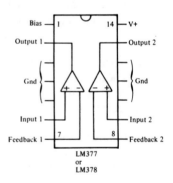

Figure 4.2   *Outline and pin notations of the LM377 dual 2 W and LM378 dual 4 W audio power amplifier ICs*

high-impedance differential input stages and fully protected output stages, and differ mainly in their voltage/power ratings and in their packaging styles. It should be noted that the input stages of these ICs are intended to be dc-biased to half-supply volts, and a bias generator is built into the ICs for this purpose.

The LM377/378/379 range of ICs is very easy to use. *Figure 4.5* shows the connections for making a simple inverting stereo amplifier powered from a single-ended power supply. Here, the amplifier is biased by connecting each

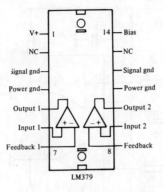

Figure 4.3   *Outline and pin notations of the LM379 dual 6 W audio power amplifier IC*

V+	LM377	LM378	LM379	Maximum output power per channel into:	
				8 R 0	16 R
12 V	Limit to 2.5 W per channel	Limit to 5 W per channel	Limit to 6 W per channel	1.6 W	1 W
16 V				2.2 W	1.5 W
18 V				3 W	1.8 W
20 V				3.8 W	2.4 W
22 V				4.6 W	2.8 W
24 V				5.4 W	3.6 W
26 V				6 W	4.2 W
28 V				7 W	5 W
30 V				–	5.5 W

Figure 4.4   *Approximate performance characteristics of the LM377/378/379 dual amplifiers*

non-inverting input pin to the bias terminal (pin-1 on the LM377 or LM378, or pin-14 on the LM379), and the closed-loop voltage gain of each amplifier is set at approximately × 50 by the ratio of R_2/R_1 or R_4/R_3. The table shows the typical performance of this circuit.

Figure 4.6 shows how the above circuit can be modified for use as a non-inverting amplifier. The voltage gain of each half is again set at roughly × 50, in this case by the ratio of R_4/R_3 or R_6/R_5, and the non-inverting input terminals are biased via the internal network of the IC.

Figure 4.7 shows how the above non-inverting amplifier circuit can be modified for use with split power supplies. Note in this case that the internal bias generator is ignored, and that the non-inverting input of each amplifier is dc-coupled to the ground half-supply point via volume control RV_1.

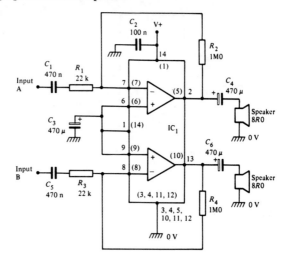

Note: LM379 pin numbers are
shown in parentheses

Typical performance of the
inverting stereo amplifier

	IC$_1$		
	LM377	LM378	LM379
V+ (max.)	18 V	24 V	28 V
P$_{out}$/CH	2 W	3 W	4 W
e$_{in}$ (max.)	80 mV	100 mV	115 mV
A$_V$ (approx.)	50	50	50
Z$_{in}$	22 k	22 k	22 k

Figure 4.5   *Simple inverting stereo amplifier using the LM377, LM378, or LM379 dual amplifier ICs*

Figure 4.8 shows a highly effective way of boosting the available output power of one half of the LM378 to 15 W via a couple of power transistors. This remarkably simple circuit generates a typical THD of only 0.05% or so at a 10 W output power level. At very low power levels, Q_1 and Q_2 are inoperative and power is fed directly to the speaker via R_2. At higher power levels Q_1 and Q_2 act as a normal complementary emitter follower and provide most of the power drive to the speaker. R_2 and the base-emitter junctions of Q_1–Q_2 are effectively wired into the negative feedback network of the circuit, thus minimizing signal crossover distortion.

Figure 4.9 shows how the above circuit can be adapted for use with a split

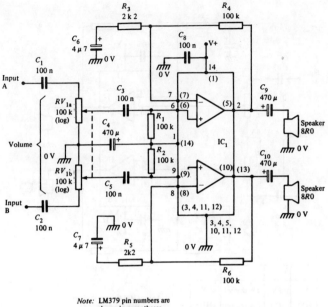

Note: LM379 pin numbers are
shown in parentheses

V+	IC_1	P_{out}
18 V	LM377	2 W/channel
24 V	LM378	3 W/channel
28 V	LM379	4 W/channel

Figure 4.6 *Non-inverting stereo amplifier using a single-ended supply*

power supply. This circuit produces negligible output dc-offset, thus enabling
the speaker to be direct-coupled to the circuit's output.

Finally, *Figure 4.10* shows how the two halves of a LM377, LM378, or
LM379 can be used to make a bridge-configured mono amplifier which can
feed relatively high power levels into a direct-coupled speaker load.

Readers considering use of the LM377 should note that NSC produce an
alternative IC that can be used in its place; this is the LM1877, a dual 2 W
power amplifier IC that is an improved pin-for-pin replacement for the
LM377 and gives a better performance in terms of very low crossover
distortion, very high input impedance, and a high slew rate, but has slightly
poorer ripple rejection and consumes a higher quiescent current than the
LM377.

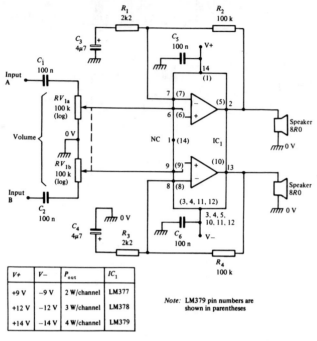

Figure 4.7 *Non-inverting stereo amplifier using a split power supply*

V+	V−	P_out	IC₁
+9 V	−9 V	2 W/channel	LM377
+12 V	−12 V	3 W/channel	LM378
+14 V	−14 V	4 W/channel	LM379

Note: LM379 pin numbers are shown in parentheses

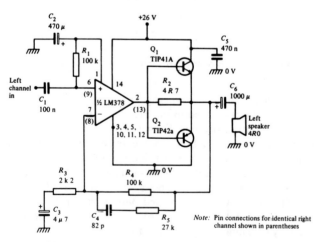

Figure 4.8 *One channel of a 15 W per channel stereo amplifier using a single-ended supply*

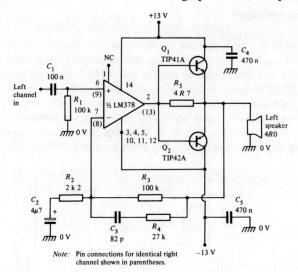

Figure 4.9 *One channel of a 15 W per channel stereo amplifier using a split supply*

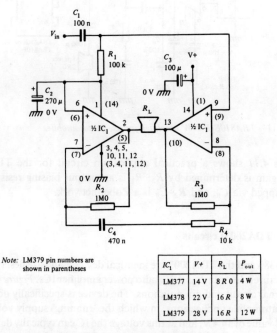

IC_1	$V+$	R_L	P_{out}
LM377	14 V	8 R 0	4 W
LM378	22 V	16 R	8 W
LM379	28 V	16 R	12 W

Note: LM379 pin numbers are shown in parentheses

Figure 4.10 *Bridge amplifier circuit using dual-amplifier ICs*

The TBA810S and P

The TBA810 is a very popular medium-power IC that is widely used in automobile applications, in which it can deliver several watts of audio power into a 4R0 load when operated from a 14.4 V (12 V nominal) supply.

The TBA810 features internal protection against accidental supply polarity inversion and high supply-line transients, etc. Early versions of this IC carry an 'S' subscript, and have a typical signal power bandwidth of only 15 kHz; recent versions carry a 'P' subscript and feature a number of performance improvements, including lower noise and a 20 kHz bandwidth.

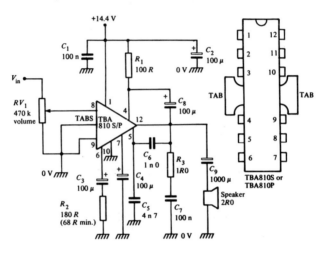

Figure 4.11   *TBA810S/P 7 W amplifier for use in automobiles*

Figure 4.11 shows a practical application circuit for the TBA810S/P; voltage gain is determined by R_2; R_1 is an output biasing resistor that is bootstrapped via C_8, and R_3–C_7 is a Zobel network.

LM383 (TDA2003) circuits

The LM383 and the TDA2003 are identical devices, and are described in the manufacturers literature as 8 W audio power amplifier ICs. *Figure 4.12* shows the internal circuit and pin notations. The device is specifically designed for use in automobile applications, in which the 'running' supply voltage has a nominal value of 14.4 V, and at this voltage the IC can typically deliver 5.5 W into a 4R0 load or 8.6 W into a 2R0 load. The IC can in fact operate with any supply voltage in the range 5 V to 20 V, can supply peak output currents of 3.5 A, and has a current-limited and thermally protected output stage.

The LM383 (TDA2003) is housed in a 5-pin package, as shown in *Figure 4.12*, and is a very easy device to use. *Figure 4.13* shows how to wire it as a

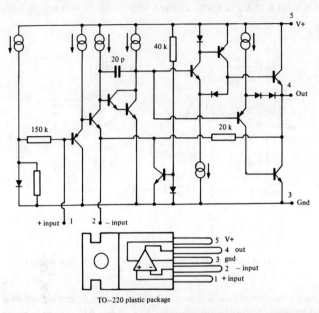

Figure 4.12 *Internal circuit and pin notations of the LM383 (TDA2003) 8 W audio power amplifier IC*

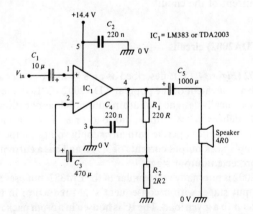

Figure 4.13 *LM383 (TDA2003) 5.5 W amplifier for use in automobiles*

5.5 W amplifier for use in automobiles. Here, the closed-loop voltage gain is set at $\times 100$ via the R_1–R_2–C_3 feedback network, and the IC is operated in the non-inverting mode by simply feeding the input signal to pin-1 via C_1. Capacitors C_2 and C_4 are used to ensure the high-frequency stability of the IC; it is vital that C_4 is wired as close as possible between pins 3 and 4.

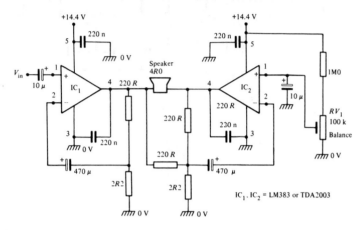

Figure 4.14   *LM383 (TDA2003) 16 W amplifier for use in automobiles*

Figure 4.14 shows how a pair of LM383 ICs can be connected as a 16 W bridge amplifier for use in automobiles. Pre-set pot RV_1 is used to balance the quiescent output voltages of the two ICs and to thus minimize the quiescent operating current of the circuit.

LM2002 (TDA2002) circuits

The LM2002 (*Figure 4.15*) is described as an 8 W audio power amplifier IC, and is a direct equivalent of the popular TDA2002 IC. Like the LM383, the LM2002 is specifically designed for use in automobile applications, in which it can typically deliver 5.2 W into a 4R0 load or 8 W into a 2R0 load. The LM2002 can in practice operate with any supply voltage in the range 5 V to 20 V, can supply peak output currents of 3.5 A, and has a current-limited and thermally protected output stage.

The LM2002 is internally very similar to the LM383, but uses a slightly less efficient output stage, with a consequent slight reduction in the available output power into a given load. The IC is housed in a 5-pin package, as shown in *Figure 4.15*, and is a very easy device to use. *Figure 4.16* shows how to wire it

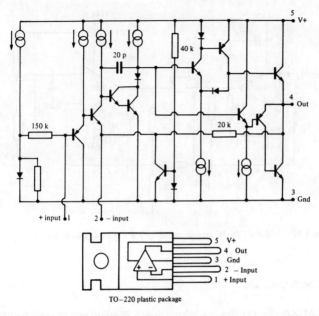

Figure 4.15 *Internal circuit and pin connections of the LM2002 (TDA2002) 8 W audio power amplifier IC*

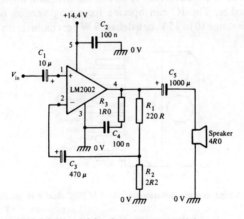

Figure 4.16 *LM2002 5.2 W amplifier for use in automobiles*

as a 5.2 W audio amplifier (with a closed-loop voltage gain set at × 100 via R_1–R_2–C_3) for use in automobiles, and *Figure 4.17* shows how to use it as a 16 W bridge amplifier for use in automobiles.

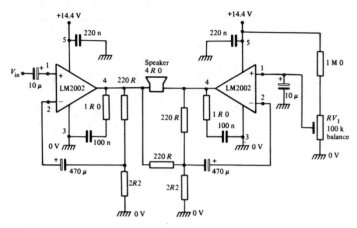

Figure 4.17   *LM2002 16 W bridge amplifier for use in automobiles*

The LM2879 circuit

The LM2879 circuit (*Figure 4.18*) is a dual 9 W audio power amplifier IC that is housed in an 11-pin package incorporating a large ground-connected metal heat tab that can be bolted directly to an external heat sink without need of an insulating washer. The IC can operate from single-ended power supply voltages in the range 10 to 35 V, can deliver 9 W per channel into an 8 R0 load,

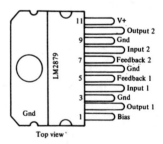

Figure 4.18   *Outline and pin notations of the LM2879 dual 9 W audio amplifier IC*

and incorporates internal current limiting and thermal shutdown circuitry, etc. It typically generates only 0.04% distortion at 4 W per channel output, and gives a 50 kHz bandwidth.

The LM2879 is a very easy device to use. *Figure 4.19* shows how it can be used as a 7 W + 7 W stereo ceramic cartridge amplifier with integral bass

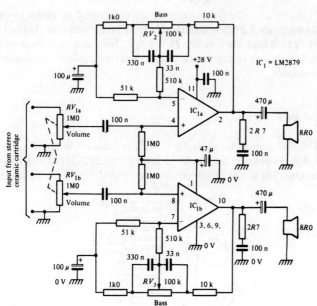

Figure 4.19   *7 W + 7 W stereo ceramic cartridge amplifier*

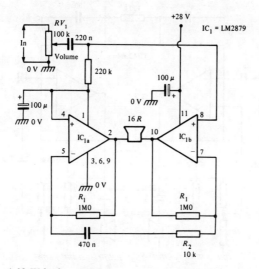

Figure 4.20   *A 12 W bridge amplifier*

controls that enable the response to be cut or boosted by up to 13 dB at 100 Hz, and *Figure 4.20* shows it connected as a bridge amplifier that can feed 12 W into a 16 *R* load when using a 28 V supply. Note in both of these circuits that pin-1 provides a voltage that is used to bias the amplifier's non-inverting input terminals.

The TDA2004 circuit

The TDA2004 circuit shown in *Figure 4.21* is a dual 9 W amplifier that is housed in an 11-pin package similar to that of the LM2879 but with different pin notations. The IC can operate from single-ended supplies in the range 8 to

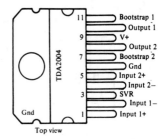

Figure 4.21   *Outline and pin notations of the TDA2004 dual 9 W audio amplifier IC*

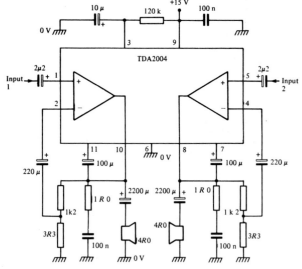

Figure 4.22   *4 W + 4 W stereo amplifier*

18 V, can provide peak output currents of 3.5 A, and can deliver 9 W into a
4R0 load from each channel when using a 17 V supply.

Figure 4.22 shows how to wire the IC as a simple stereo amplifier that will
deliver 4 W per channel into 4R0 loads while powered from a 15 V supply and
generating total harmonic distortion of less than 0.2%. Note that each
channel has its voltage gain set at about × 364 by the ratio of the 1k2 and 3R3
resistors, and each channel thus needs an input of only 12 mV rms to give full
output.

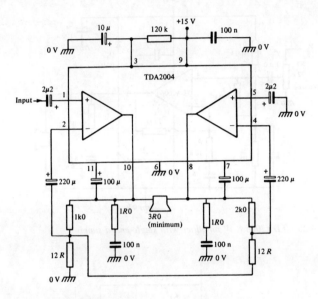

Figure 4.23   *A 20 W bridge amplifer*

Figure 4.23 shows how to wire the TDA2004 as a bridge amplifier that can
deliver a maximum of about 20 W into a 3R0 load (the minimum allowable
value) when using a 15 V supply. This circuit needs an input of about 50 mV
rms for full output.

The TDA2006 circuit

This is a high-quality amplifier that can be used with either split or single-
ended power supplies and can deliver up to 12 W into a 4R0 load, and which
typically generates less than 0.1% distortion when feeding 8 W into a 4R0
speaker. The IC is housed in a 5-pin TO220 package (see *Figure 4.24*) that has

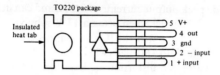

Figure 4.24   *Outline and pin connections of the TDA2006 and TDA2030*

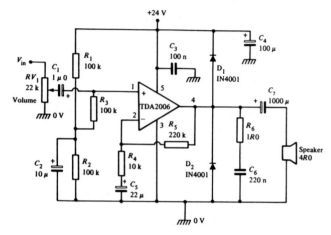

Figure 4.25   *TDA2006 8 W amplifier with single-ended supply*

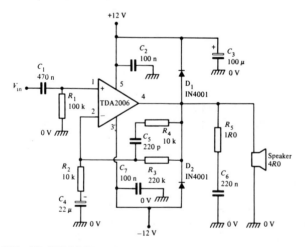

Figure 4.26   *TDA2006 8 W amplifier with split power supply*

an electrically insulated heat tab which can be bolted directly to an external heat sink without need of an insulating washer.

Figure 4.25 shows how to use the TDA2006 with a single-ended supply. The non-inverting input pin is biased at half-supply volts via R_3 and the R_1–R_2 potential divider, and the voltage gain is set at $\times 22$ via R_5/R_4. D_1 and D_2 protect the output of the IC against damage from back electromotive force (EMF) voltages from the speaker, and R_6–C_6 form a Zobel network.

Figure 4.26 shows how to modify the above circuit for use with split power supplies. In this case the non-inverting input is tied to ground via R_1. This circuit also shows how high-frequency roll-off can be applied to the amplifier via C_5–R_4.

The TDA2030 circuit

This very popular IC can be regarded as an up-rated version of the TDA2006, and is housed in the same 5-pin TO220 package with insulated heat tab. It can operate with single-ended supplies of up to 36 V, or with balanced split supplies of up to 18 V. When used with a +28 V single-ended supply it gives a guaranteed output of 12 W into 4R0 or 8 W into 8R0. Typical THD is 0.05% at 1 kHz at 7 W output, and is still less than 0.1% at 8 W.

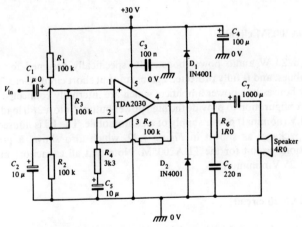

Figure 4.27   *TDA2030 15 W amplifier with single-ended supply*

Figure 4.27 shows how to connect the TDA2030 as a 15 W amplifier using a single-ended +30 V supply and a 4R0 speaker load and a voltage gain of 30 dB. Alternatively, *Figure 4.28* shows how to wire a pair of these ICs as a split-supply bridge amplifier that can deliver 24 W into a 4R0 speaker load while generating typical total harmonic distortion of less than 0.5%.

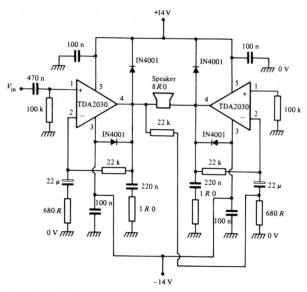

Figure 4.28   *TDA2030 24 W bridge amplifier with split supply*

The TDA2005M circuit

This is a 20 W audio power booster IC specifically designed for use in automobiles, and is fully protected against output short circuits, etc. The IC actually houses two power amplifiers which are internally connected in the bridge configuration to provide the high power output (into a 2R0 load) from the 14.4 V (nominal) power supply of an automobile. The IC is housed in an 11-pin package, as shown in *Figure 4.29*, which also shows a practical applications circuit for the TDA2005M. Note that all capacitors must be rated at 25 V minimum.

The TDA1520 circuit

This is a very high performance device that can deliver up to 22 W into a 4R0 load or 11 W into an 8R0 load when powered from a 33 V supply, and which typically generates a mere 0.01% distortion at 16 W output into a 4R0 load. The IC is housed in a 9-pin package that can be bolted directly to an external heat sink (without the need for insulating washers) in single-ended supply applications. *Figure 4.30* shows the IC outline and a practical application circuit for this device.

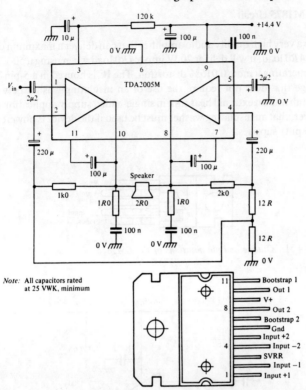

Figure 4.29   *TDA2005M 20 W power booster for use in cars*

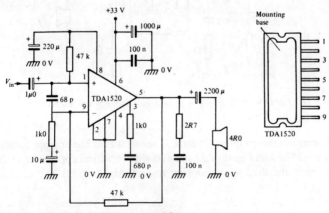

Figure 4.30   *TDA1520 22 W power amplifier*

The LM1875 circuit

This is a very high quality audio amplifier that can deliver a maximum of 25 W into a 4R0 load; it will deliver 20 W into a 4R0 load when using a 50 V supply and generating a mere 0.015% distortion. The IC is housed in a 5-pin TO220 package that does not require the use of an insulating washer between its metal tab and an external heat sink in single-ended supply applications; note, however, that an insulating washer must be used if the device is powered from dual (split) supplies.

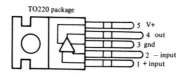

Figure 4.31   *Outline and pin notations of the LM1875*

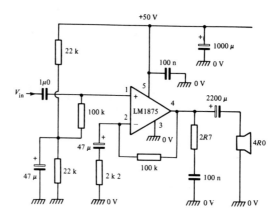

Figure 4.32   *LM1875 20 W amplifier using single-ended power supply*

To complete this chapter, *Figure 4.31* shows the outline and pin notations of the LM1875, and *Figures 4.32* and *4.33* show practical application circuits using single and dual power supplies respectively.

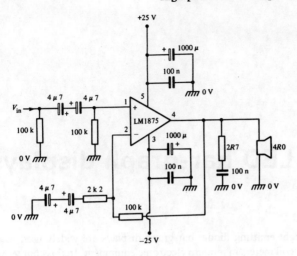

Figure 4.33 *LM1875 20 W amplifier using dual (split) power supply*

5 LED bar-graph displays

LED (light emitting diode) bar-graph displays are widely used to replace moving-coil meters in modern electronic equipment. In this chapter we look at two distinct families of bar-graph driver ICs – the U237 type from AEG and the LM3914 type from NSC – and show how to use these devices.

LED bar-graph basics

LED bar-graph displays can be regarded as modern all-electronic moving-light replacements for the old-fashioned moving coil-and-pointer types of analogue indicating meter, and are widely used in modern domestic equipment such as amplifiers and disc recorder units, etc. In a bar-graph display a line of ordinary LEDs are used to give an analogue representation of a scale length, and in use several adjacent LEDs may be illuminated simultaneously to form a light bar that gives an analogue indication of the value of a measured parameter.

Figure 5.1 illustrates the bar-graph indicating principle, and shows a line of

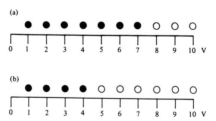

Figure 5.1   *(a) Bar-graph indication of 7 V on a 10 V 10-LED scale, and (b) bar-graph indication of 4 V on a 10-V 10-LED scale*

112

ten LEDs used to represent a linear-scale 0-to-10 V meter. *Figure 5.1(a)* shows the display indicating an input of 7 V, with 7 LEDs on and 3 LEDs off. *Figure 5.1(b)* shows the display indicating an input of 4 V, with 4 LEDs on and 6 LEDs off.

A number of special bar-graph driver ICs are available for activating LED bar-graph displays. The two most useful types are the U237, etc. family from AEG, and the LM3914, etc. family from NSC. The U237 family are simple 'dedicated' devices which can usefully be cascaded to drive a maximum of 10 LEDs in bar mode only. The LM3914 family are more complex and highly versatile devices, which can usefully be cascaded to drive as many as 100 LEDs, and can drive them in either bar or dot mode. *Figure 5.2(a)* shows a 10-LED 10 V meter indicating 7 V in the 'dot' mode, and *Figure 5.2(b)* shows the same meter indicating an input of 4 V. Note in each case that only a single LED is illuminated, and that its scale position indicates the analogue value of the input voltage.

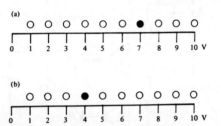

Figure 5.2   *(a) Dot indication of 7 V on a 10 V 10-LED scale, and (b) dot indication of 4 V on a 10-V 10-LED scale*

Bar-graph displays (with suitable drivers) act as inexpensive and superior alternatives to analogue-indicating moving-coil meters. They are immune to inertia and sticking problems, so are fast acting and are unaffected by vibration or attitude. Their scales can easily be given any desired shape (a vertical or horizontal straight line, an arc or circle, etc.). In a given display, individual LED colours can be mixed to emphasize particular sections of the display. Electronic over-range detectors can easily be activated from the driver ICs and used to sound an alarm and/or flash the entire display under the over-range condition.

Bar-graph displays have far better linearity than conventional moving-coil meters, typical linear accuracy being 0.5%. Scale definition depends on the number of LEDs used; a 10-LED display gives adequate resolution for most practical purposes.

Let us now move on and take detailed looks at the two main families of bar-graph driver ICs, starting off with the U237 types.

U237 bar-graph driver ICs

Basic principles

The U237 family of bar-graph driver ICs are manufactured by AEG. They are simple, dedicated devices, housed in 8-pin DIL packages and each capable of directly driving up to five LEDs. The family comprises four individual devices. The U237B and U247B produce a linear-scaled display and are intended to be used as a pair, driving a total of ten LEDs. The U257B and U267B produce a log-scaled display and are also intended to be used as a pair driving a total of ten LEDs.

All ICs of the U237 family use the same basic internal circuitry, which is shown in block diagram form (together with external connections) in *Figure*

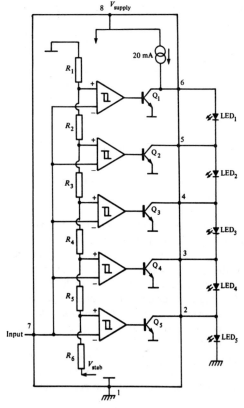

Figure 5.3 *Block diagram of the U237-type bar-graph driver, with basic external connections*

5.3. The IC houses five sets of Schmitt voltage-comparators/transistor-switches, each of which has its threshold switching or step voltage individually determined by a tapping point on the R_1 to R_5 voltage divider, which is powered from a built-in voltage regulator: the input of each comparator is connected to the pin-7 input terminal of the IC. The IC also houses a constant-current generator (20 mA nominal), and the five external LEDs are wired in series between this generator and ground (pin-1), as shown in *Figure 5.3*.

The basic action of the circuit is such that groups of LEDs are turned on or off by activating individual switching-transistors within the IC. Thus, if Q_3 is turned on it sinks the 20 mA constant-current via LEDs 1 and 2, so LEDs 1 and 2 turn on and LEDs 3 to 5 turn off.

The U237B has step voltages spaced at 200 mV intervals, and *Figure 5.4* shows the states of its five internal transistors at various values of input voltage. Thus, at zero volts input, all five transistors are switched on, so Q_1 sinks the full 20 mA of constant-current, and all five LEDs are off. At 200 mV input, Q_1 turns off but all other transistors are on, so Q_2 sinks the 20 mA via

V_{in}	Q_1	Q_2	Q_3	Q_4	Q_5
1.0 V	Off	Off	Off	Off	Off
0.8 V	Off	Off	Off	Off	On
0.6 V	Off	Off	Off	On	On
0.4 V	Off	Off	On	On	On
0.2 V	Off	On	On	On	On
0 V	On	On	On	On	On

Figure 5.4 *States of the U237B internal transistors at various input voltages*

LED$_1$, driving LED$_1$ on and causing all other LEDs to turn off, and so on. Eventually, at 1 V input, all transistors are off and the 20 mA flows to ground via all LEDs, so all five LEDs are on. Note that the operating current of the circuit is virtually independent of the number of LEDs turned on, so the IC generates negligible radio frequency interference as it switches transistors/LEDs.

Device	Step 1	Step 2	Step 3	Step 4	Step 5
U237 B	200 mV	400 mV	600 mV	800 mV	1.00 V
U247 B	100 mV	300 mV	500 mV	700 mV	900 mV
U257 B	0.18 V/−15 dB	0.5 V/−6 dB	0.84 V/−1.5 dB	1.19 V/+1.5 dB	2.0 V/+6 dB
U267 B	0.1 V/−20 dB	0.32 V/−10 dB	0.71 V/−3 dB	1.0 V/0 dB	1.41 V/+3 dB

Figure 5.5 *Step voltage values of the U237 family of bar-graph driver ICs*

The four ICs in the U237 family differ only in their values of step voltages, which are determined by the R_1 to R_5 potential divider values. *Figure 5.5* shows the step values of the four ICs. Note that the U237B and U247B are linearly scaled, and can be coupled together to make a 10-LED linear meter with a basic full-scale value of 1 V. The U257B and U267B are log scaled, and can be coupled together to make a 10-LED log meter with a basic full-scale value of 2 V or +6 dB.

What supply voltage?

Figure 5.6 shows the basic specification of the U237 family of ICs. Note that the supply voltage range is specified as 8 to 25 V. In practice, the minimum supply voltage is one of the few design points that must be considered when using these devices, and must be at least equal to the sum of the ON voltages of the five LEDs, plus a couple of volts to allow correct operation of the internal constant-current generator. Thus, when driving five red LEDs, each with a forward volt drop of 2 V, the supply value must be at least 12 V. Different coloured LEDs, with different forward volt drop values, can be used together in the circuit, provided that the supply voltage is adequate.

Parameter	Minimum	Typical	Maximum
Supply voltage (see text)	8 V	12 V	25 V
Input voltage			5 V
Input current			0.5 mA
Maximum supply current		25 mA	30 mA
Power dissipation (at 60°C)			690 mW
Step tolerance	−30 mV		+30 mV
Step hysteresis		5 mV	10 mV
Input resistance		100 k	
Output saturation voltage			1 V

Figure 5.6   *General specification of the U237 family of ICs*

The only other 'usage' point concerns the input impedance of the IC. Although the input impedance is high (typically greater than 100 k), the IC in fact tends to become unstable if fed from a source impedance in excess of 20 k or so. Ideally, the signal feeding the input should have a source impedance less than 10 k. If the source impedance is greater than 10 k, stability can be enhanced by wiring a 10 n capacitor between pins 7 and 1.

Practical U237 circuits

Figures 5.7 to *5.14* show some practical ways of using the U237 family of devices. In all of these diagrams we have shown the supply voltages as being +12 to 25 V, but the reader should keep in mind the constraints already mentioned.

Figure 5.7 shows the practical connections for making a 0–1 V 5-LED linear-scaled meter, using a single U237B IC, and *Figure 5.8* shows how a U237B/U247B pair of ICs can be coupled together to make a 0–1 V 10-LED

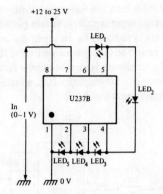

Figure 5.7 *Practical connections for making a 0–1 V 5-LED linear-scaled meter*

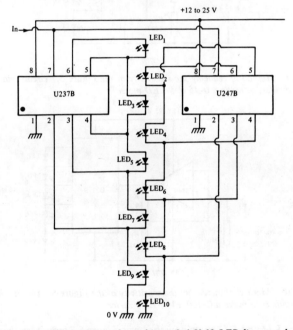

Figure 5.8 *Practical connections for making a 0–1 V 10-LED linear-scaled meter*

linear-scaled meter. Note in the latter case that the two ICs are operated as individual *Figure 5.7* circuits (needing only a 5-LED supply voltage), but have their input terminals tied together and have their LEDs *physically* alternated, to give a 10-LED display.

Figures 5.9 and *5.10* show how the full-scale sensitivity of the basic circuit can be altered, via external input circuitry, to suit particular applications. In *Figure 5.9*, the sensitivity is reduced by an input potential divider ($R_1-R_2-RV_1$), which has a 15:1 ratio and thus gives an effective full-scale sensitivity of 15 V. In *Figure 5.10*, the sensitivity is increased by a factor of ten, to 100 mV full-scale, via non-inverting × 10 amplifier IC_2, which also raises the input impedance to *1M0* (determined by R_1).

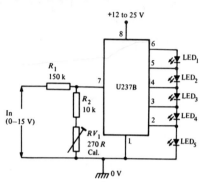

Figure 5.9   *Method of reducing the sensitivity of the* Figure 5.7 *circuit, via an input potential divider, to make a 0–15 V 5-LED meter*

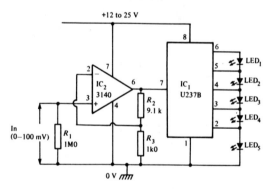

Figure 5.10   *Method of increasing the sensitivity of the* Figure 5.7 *circuit, via a × 10 buffer amplifier, to make a 0–100 mV 5-LED meter*

Figures 5.11 and *5.12* show how the basic *Figure 5.7* circuit can be used to indicate the value of a physical parameter, such as light, heat, liquid level, etc., that can be represented by an analogue resistive value in a transducer (R_T). In both of these circuits the transducer is simply fed from a constant-current generator, so that the input voltage reaching the IC is directly proportional to the transducer resistance.

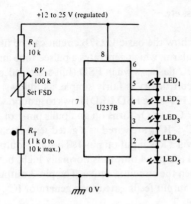

Figure 5.11 *Simple method of using a transducer sensor to indicate the value of a physical quantity*

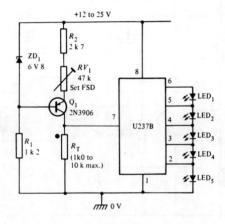

Figure 5.12 *Alternative method of using a transducer sensor to indicate the value of a physical quantity*

In the *Figure 5.11* circuit, the constant current is derived from the regulated supply line via R_1–RV_1, and current constancy relies on the fact that the supply voltage is large relative to the 1 V full-scale value of the meter. Thus, if the supply value is 20 V, the transducer current varies by only 5% when the transducer resistance varies between the zero volts and full-scale volts values. *Figure 5.12* shows how the linearity can be further improved, without the need for a regulated supply, by feeding the transducer from the output of the Q_1 constant-current generator.

Over-range alarms, etc.

Figure 5.13 shows how the basic U237B circuit can be fitted with an audio-visual over-range alarm, which generates a pulsed tone and flashes the entire display at a rate of 2 flashes/s when FSD (full-scale deflection) is reached or exceeded. The circuit theory is fairly simple, as follows:

The current of LED_5 (the FSD LED) flows to ground via R_1 and the base-emitter junction of Q_1, so Q_1 turns on and pulls pin-1 of IC_{2a} low whenever LED_5 turns on. IC_{2a}–IC_{2b} are wired as a gated semi-latching 0.5 Hz astable, which is gated on via a low input on pin-1. The pin-3 output of this astable is normally low, and the pin-4 output is normally high, but both pins give an astable output when the circuit is active. The pin-3 output feeds the base of Q_2, and the pin-4 output feeds gated tone generator IC_{2c}–IC_{2d}, which feeds acoustic transducer T_x.

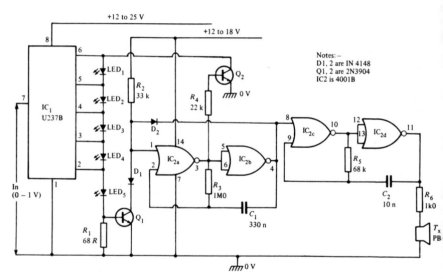

Figure 5.13   *Method of fitting an audio-visual over-range alarm to the basic* Figure 5.7 *circuit; the entire display flashes (at a rate of 2 flashes/s) when FSD is reached*

Thus, when IC_1 output is below FSD, Q_1 is off and Q_2 and the tone generator are inactive. When FSD is reached, LED_5 and Q_1 turn on, and pin-1 of IC_{2a} is pulled low via D_1, and IC_{2a}–IC_{2b} enters an astable mode in which pin-4 switches low and activates the IC_{2c}–IC_{2d} acoustic alarm and also pulls pin-1 low via D_2; simultaneously, pin-3 switches high and turns Q_2 on, thereby sinking the entire 20 mA constant current of IC_1 and blanking all LEDs for half the duration of the astable cycle. Note that Q_1 turns off as soon as Q_1 turns on, but that the input (pin-1) of the astable is maintained low at this stage via D_2. At the end of the half-cycle, Q_2 turns off and restores the display and the IC_{2c}–IC_{2d} acoustic alarm turns off. As the display is restored, Q_1 turns back on (if LED_5 is still active), but the IC_{2a}–IC_{2b} astable cannot re-trigger until it completes the second 'free-wheeling' half of its cycle.

Thus, the *Figure 5.13* circuit flashes the entire display and generates a pulsed-tone alarm under the FSD or over-range condition. Note that if the IC_1 supply rail is restricted to the 12-to-18 V range, it can be made common with the supply of the alarm circuitry. If the acoustic alarm is not required, simply omit R_5–C_2–R_6–T_x and connect the inputs of IC_{2c} and IC_{2d} to ground. The circuit can be made to flash only part of the display, if required, by simply connecting Q_2 collector to the appropriate one of the pin-2-to-6 terminals of IC_1; connection to pin-3, for example, causes only LEDs 4 and 5 to flash. If the over-range alarm circuit is to be used with the *Figure 5.8* 10-LED circuit, the feed to R_1 and the base of Q_1 should be taken from LED_{10}.

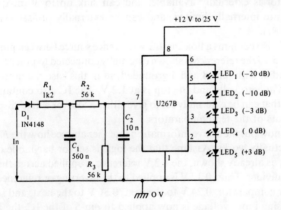

Figure 5.14   *5-LED AF-level meter. A 10-LED version can be made by using a U257B/U267B pair of ICs*

Finally, to complete this look at the U237 range of ICs, *Figure 5.14* shows how the U267B 'log' IC can be used to make a 5-LED audio-level meter. A 10-LED meter can be made by connecting the R_1–R_2–R_3–C_1–D_1 input circuit to the input of a U257B/U267B pair of ICs connected in the same configuration as shown in *Figure 5.8*.

The basic *Figure 5.14* circuit has a discharge time constant of about 70 ms; the sensitivity is determined by the R_2–R_3 ratio and, with the values shown, indicates 0 dB with an input of 3 V. The circuit requires a low-impedance drive, such as can be obtained (directly or via a potential divider) from a loud-speaker, etc.

LM3914 dot/bar-graph driver ICs

Basic principles

The LM3914 family of ICs are manufactured by NSC. They are fairly complex and highly versatile devices, housed in 14-pin DIL packages and each capable of directly driving up to ten LEDs in either dot or bar mode. The family comprises three devices, the LM3914 being a linear-scaled unit and the LM3915 and LM3916 being log and semi-log devices respectively.

All three devices in the LM3914 family use the same basic internal circuitry, and *Figure 5.15* shows the specific internal circuit of the linear-scaled LM3914, together with the connections for making it act as a simple 10-LED 0–1.2 V meter. The IC contains ten voltage comparators, each with its non-inverting terminal taken to a specific tap on a 'floating' precision multi-stage potential divider and with all inverting terminals wired in parallel and taken to input pin-5 via a unity-gain buffer amplifier. The output of each comparator is externally available, and can sink up to 30 mA; the sink currents are internally limited, and can be externally pre-set via a single resistor (R_1).

The IC also contains a floating 1.2 V reference source between pins 7 and 8. In *Figure 5.15* the reference is shown externally connected to potential divider pins 6 and 4, with pins 8 and 4 grounded, so in this case the bottom of the divider is at zero volts and the top is at 1.2 V. The IC also contains a logic network that can be externally set to give either a 'dot' or a 'bar' display from the outputs of the ten comparators.

Let us now put the above information together and see how the *Figure 5.15* circuit actually works. Assume that the logic is set for bar-mode operation and that, as already shown, the 1.2 V reference is applied across the internal 10-stage divider. Thus, 0.12 V is applied to the inverting or reference input of the lower comparator, 0.24 V to the next, 0.36 V to the next, and so on. If a slowly rising input voltage is now applied to pin-5 of the IC, the following sequence of actions takes place:

When the input voltage is zero, the outputs of all ten comparators are disabled and all LEDs are off. When the input voltage reaches the 0.12 V reference value of the first comparator, its output conducts and turns LED_1 on. When the input reaches the 0.24 V reference value of the second comparator, its output also conducts and turns on LED_2, so at this stage LED_1 and LED_2 are both on. As the input voltage is further increased, progressively more and more comparators and LEDs are turned on, until eventually, when the input rises to 1.2 V, the last comparator and LED_{10} turn on, at which point all ten LEDs are illuminated.

A similar kind of action is obtained when the LM3914 logic is set for dot mode operation, except that only one LED turns on at any given time. At zero volts, no LEDs are on, and at above 1.2 V only LED_{10} is on.

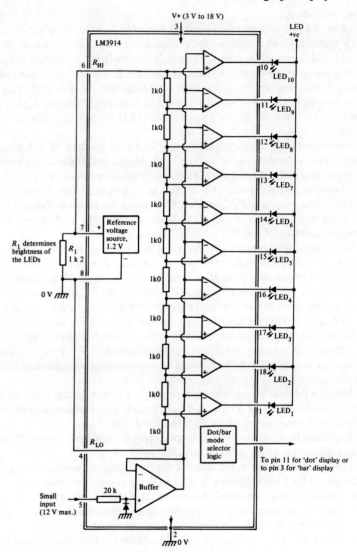

Figure 5.15 *Internal circuit of the LM3914, with connections for making a 10-LED 0–1.2 V linear meter with dot or bar display*

Some finer details

In *Figure 5.15*, R_1 is shown connected between pins 7 and 8 (the output of the 1.2 V reference), and determines the on currents of the LEDs. The on current of each LED is roughly ten times the output current of the 1.2 V source, which can supply up to 3 mA and thus enables LED currents of up to 30 mA to be set via R_1. If, for example, a total resistance of 1k2 (equal to the paralleled values of R_1 and the 10 k of the IC's internal potential divider) is placed across pins 7 and 8, the 1.2 V source will pass 1 mA and each LED will pass 10 mA in the on mode.

Note from the above that the IC can pass total currents up to 300 mA when used in the 'bar' mode with all ten LEDs on. The IC has a maximum power rating of only 660 mW, so there is a danger of exceeding this rating when the IC is used in the bar mode. In practice, the IC can be powered from DC supplies in the range 3 to 25 V, and the LEDs can use the same supply as the IC or can be independently powered; this latter option can be used to keep the power dissipation of the IC at minimal level.

The internal 10-stage potential divider of the IC is floating, with both ends externally available for maximum versatility, and can be powered from either the internal reference or from an external source or sources. If, for example, the top of the chain is connected to a 10 V source, the IC will function as a 0–10 V meter if the low end of the chain is grounded, or as a 'restricted-range' 5–10 V meter if the low end of the chain is connected to a 5 V source. The only constraint on using the divider is that its voltage must not be greater than 2 V less than the IC's supply voltage (which is limited to 25 V maximum). The input (pin-5) to the IC is fully protected against overload voltages up to plus or minus 35 V.

The IC's internal voltage reference produces a nominal output of 1.28 V (limits are 1.2 V to 1.32 V), but can be externally programmed to produce effective reference values up to 12 V (we show how shortly).

The IC can be made to give a dot-mode display by wiring pin-9 to pin-11, or a bar display by wiring pin-9 to positive-supply pin-3.

Finally, it should be noted that the major difference between the three members of the LM3914 family of ICs lays in the values of resistance used in the internal 10-stage potential divider. In the LM3914, all resistors in the chain have equal values, and thus produce a linear display of ten equal steps. In the LM3915, the resistors are logarithmically weighted, and thus produce a log display that spans 30 dB in ten 3 dB steps. In the LM3916, the resistors are weighted in a semi-log fashion and produce a display that is specifically suited to vu-meter applications.

Let us now move on and look at some practical applications of this series of devices, paying particular attention to the linear LM3914 IC.

Practical LM3914 circuits

Dot-mode voltmeters
Figures 5.16 to *5.19* show various ways of using the LM3914 IC to make 10-LED dot mode voltmeters. Note in all of these circuits that pin-9 is wired to pin-11, to give dot-mode operation, and that a 10 μ capacitor is wired directly between pins 2 and 3 to enhance circuit stability.

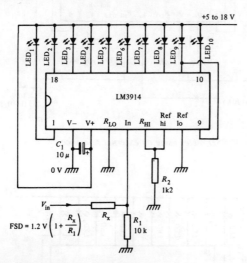

Figure 5.16   *1.2–1000 V FSD dot-mode voltmeter*

Figure 5.16 shows the connections for making a variable-range (1.2 V to 1000 V FSD) voltmeter. The low ends of the internal reference and divider are grounded and their top ends are joined together, so the meter has a basic full-scale sensitivity of 1.2 V, but variable ranging is provided by the R_x–R_1 potential divider at the input of the circuit. Thus, when R_x is zero, FSD is 1.2 V, but when R_x is 90 k the FSD is 12 V. Resistor R_2 is wired across the internal reference and sets the on current of all LEDs at about 10 mA.

Figure 5.17 shows how to make a fixed-range 0–10 V meter, using an external 10 V zener diode (connected to the top of the internal divider) to provide a reference voltage. The supply voltage to this circuit must be at least 2 V greater than the zener reference voltage.

Figure 5.18 shows how the internal reference of the IC can be made to effectively provide a variable voltage, enabling the meter FSD value to be set anywhere in the range 1.2 V to 10 V. In this case the 1 mA current (determined by R_1) of the floating 1.2 V internal reference flows to ground via RV_1, and the resulting RV_1-voltage raises the reference pins (7 and 8) above zero. If, for

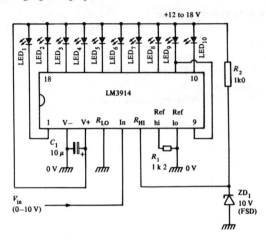

Figure 5.17   *10 V FSD meter using an external reference*

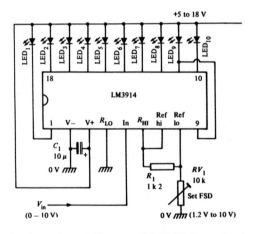

Figure 5.18   *An alternative variable-range (1.2–10 V) dot-mode voltmeter*

example, RV_1 is set to 2k4, pin-8 will be at 2.4 V and pin-7 at 3.6 V. RV_1 thus enables the pin-7 voltage, which is connected to the top of the internal divider, to be varied from 1.2 V to approximately 10 V, and thus sets the FSD value of the meter within these values.

Finally, *Figure 5.19* shows the connections for making an expanded-scale meter that, for example, reads voltages in the range 10–15 V. RV_2 sets the

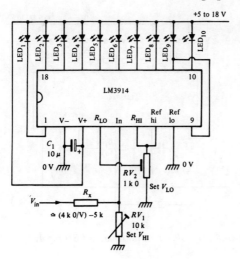

Figure 5.19 *Expanded-scale (10–15 V) dot-mode voltmeter*

LED current at about 12 mA, but also enables a reference value in the range 0–1.2 V to be set on the low end of the internal divider. Thus, if RV_2 is set to apply 0.8 V to pin-4, the basic meter will read voltages in the range 0.8–1.2 V only. By fitting potential divider R_x–RV_1 to the input of the circuit, this range can be 'amplified' to, say, 10–15 V, or whatever range is desired.

Bar-mode voltmeters

The dot-mode circuits of *Figures 5.16* to *5.19* can be made to give bar-mode operation by simply connecting pin-9 to pin-3, rather than to pin-11. When using the bar mode, however, it must be remembered that the IC's power rating must not be exceeded by allowing excessive output-terminal voltages to be developed when all ten LEDs are on. LEDs 'drop' about 2 V when they are conducting, so one way around this problem is to power the LEDs from their own low-voltage (3 to 5 V) supply, as shown in *Figure 5.20*.

An alternative solution is to power the IC and the LEDs from the same supply, but to wire a current-limiting resistor in series with each LED, as shown in *Figure 5.21*, so that the IC's output terminals saturate when the LEDs are on.

An alternative way of obtaining a bar display without excessive power dissipation is shown in *Figure 5.22*. Here, the LEDs are all wired in series, but with each one connected to an individual output of the IC, and the IC is wired for dot-mode operation. Thus, when, for example, LED_5 is driven on it draws its current via LEDs 1 to 4, so all five LEDs are on. In this case, however, the

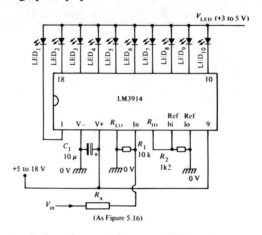

Figure 5.20   *Bar-display voltmeter with separate LED supply*

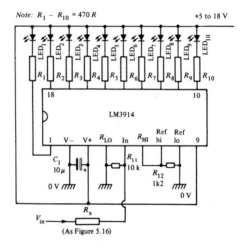

Figure 5.21   *Bar-display voltmeter with common LED supply*

total LED current is equal to that of a single LED, so power dissipation is quite low: in this mode the LM3914 in fact operates in a similar fashion to the U237-type of IC. The LED supply to this circuit must be greater than the sum of all LED volt drops when all LEDs are on, but within the voltage limits of the IC; a regulated 24 V supply is thus needed.

Figure 5.23 shows a modification of the above circuit, which enables it to be powered from an unregulated 12–18 V supply. In this case the LEDs are split into two chains, and the transistors are used to switch the lower (LEDs 1 to 5) chain on when the upper chain is active; the maximum total LED current is equal to twice the current of a single LED.

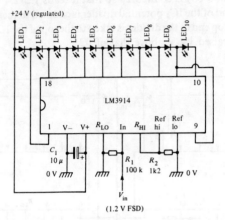

Figure 5.22   *Method of obtaining a bar display with dot-mode operation and minimal current consumption*

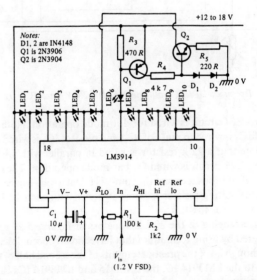

Figure 5.23   *Modification of the* Figure 5.22 *circuit, for operation from unregulated 12–18 V supplies*

20-LED voltmeters

To complete this chapter, *Figures 5.24* and *5.25* show how pairs of LM3914s can be interconnected to make 20-LED 0–2.4 V voltmeters. Here, the input terminals of the two ICs are wired in parallel, but IC_1 is configured so that it reads 0–1.2 V and IC_2 is configured so that it reads 1.2–2.4 V. In the latter case, the low end of the IC_2 potential divider is coupled to the 1.2 V reference of IC_1 and the top end of the divider is taken to the top of the 1.2 V reference of IC_2, which is raised 1.2 V above that of IC_1.

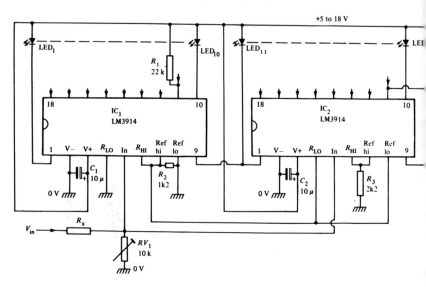

Figure 5.24   *Dot-mode 20-LED voltmeter (FSD = 2.4 V when $R_x = 0$)*

The *Figure 5.24* circuit is wired for dot-mode operation. Note in this case that pin-9 of IC_1 is wired to pin-1 of IC_2, and pin-9 of IC_2 is wired to pin-11 of IC_2. Also note that a 22 k resistor is wired in parallel with LED_9 of IC_1.

The *Figure 5.25* circuit is wired for bar-mode operation. The connections are similar to those above, except that pin-9 is taken to pin-3 of each IC, and a 470 R current-limiting resistor is wired in series with each LED to reduce power dissipation of the ICs.

The voltage-ranging of the 20-LED meters of *Figures 5.24* and *5.25* can easily be altered by using techniques that have already been described. Also, note that although all of the practical circuits of the second half of this chapter are devoted to the LM3914 IC, the LM3915 and LM3916 ICs can in fact be directly substituted in most of these circuits, to give log and semi-log displays respectively.

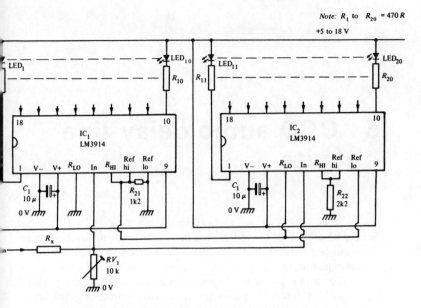

Figure 5.25   *Bar-mode 20-LED voltmeter (FSD = 2.4 V when* $R_x = 0$*)*

Finally, note that an over-range alarm of the type shown in *Figure 5.13* can be fitted to an LM3914-type IC by wiring a transistor in series with the top LED to detect the full-scale state, and by wiring a transistor switch in series with the LED-brightness resistor to pulse the display on and off under the over-range condition.

6 CCD audio delay-line circuits

Solid state delay lines are widely used in modern music and audio systems. They can be used to produce popular sound effects such as echo, reverb, chorus or phasing in music systems, or rare effects such as ambience synthesis or pseudo room expansion in expensive hi-fi systems, or to give predictive effects such as automatic pre-switching in tape recorders or click/scratch elimination in record players, or to equalize speech delays in public address systems, etc.

Solid state delay-line systems are available in both analogue and digital forms; analogue types are currently the most popular, and several companies manufacture dedicated analogue delay lines in linear IC form. In this chapter we discuss the operating principles of such ICs, and show how they can be used in practical delay-line systems.

Delay-line basics

Analogue delay-line ICs are widely available and are generally know as charge coupled device (CCD) or bucket brigade delay lines. In essence, they each contain a chain of hundreds of series-connected sample-and-hold analogue memory cells or 'buckets' (hence the bucket brigade title) that are clocked via an external generator so that the device acts like an analogue shift register; analogue input signals are applied at the front of the bucket chain, and are clocked out in time-delayed form at the chain's end.

Figure 6.1 illustrates the basic operating principles of an analogue delay line. Each memory cell (bucket) consists of a small capacitor and a tetrode MOSFET and acts like a sample-and-hold stage. An electronic switch (SW_1) is placed at the front of the chain, which is externally biased to a pre-set voltage. Charges can be shifted down the chain, one step at a time, via an

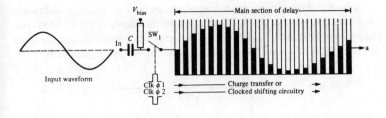

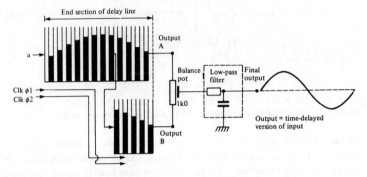

Figure 6.1 *Basic operating principle of the bucket brigade delay line. See text for explanation*

external 2-phase clock, which is also used to activate input sampling switch SW_1. The operating sequence is as follows.

On the first clock half-cycle, each existing bucket charge is shifted backwards one step, to the next bucket in the chain, and a sample of the instantaneous input signal voltage is fed (via SW_1) to the first bucket and stored as an analogue charge. On the second half cycle, each existing charge (including the input one) is transferred backwards another step, but the input is NOT sampled via SW_1. There is thus always an 'empty' bucket between each charged bucket in the chain. This double shifting process repeats on each clock cycle, with input samples repeatedly being taken and then clocked backwards one step.

In the final part of the delay line a short bucket section is wired in parallel with, and fed from, the main delay line, but has one bucket more than the corresponding section of the main line and is clocked in anti-phase. The IC thus has two outputs which, when added together, effectively fill in the 'gaps' in the main delay line bucket chain. The outputs can be 'added' either by shorting them directly together or, preferably, by connecting them to a balance pot as shown in the diagram. The final output of the delay line is thus a quantized but time-delayed replica of the original input signal.

Figure 6.2 shows the essential 'usage' elements of an analogue delay-line chip. The delay-line MOSFETs use a tetrode structure, so the IC needs two supply lines (V_{DD} and V_{BB} or V_{GG}), plus a ground connection, and the input terminal must be biased into the linear mode by voltage V_{bias}. The IC's two

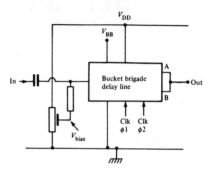

Figure 6.2   *Essential 'usage' elements of an analogue delay-line chip*

outputs must be added together, as already described; in *Figure 6.2* we have shown addition by direct-shorting. Finally, the IC needs a 2-phase clock signal, normally consisting of a pair of anti-phase square waves that switch fully between the V_{DD} and gnd potentials.

How much delay?

We have seen that delay-line buckets are alternately 'empty' and 'charged', and that each complete clock cycle shifts a charge two stages along the bucket chain. Thus, the maximum number of samples taken by a line is equal to half the number of bucket stages, and the actual time-delay available from a line is given by:

Time delay $= S.p/2$ or $S/2.f$,

where S is the number of bucket stages, p is the clock period, and f is the clock frequency.

Practical analogue delay-line ICs usually have 512, 1024, 1536 or 4096 stages. Thus, a 1024-stage line using a 10 kHz (100 μs) clock gives a delay of 51.2 ms. A 4096-stage line gives a 204.8 ms delay at the same clock frequency. Note that in practice the maximum useful signal frequency of a delay line is equal to one third of the clock frequency, so a delay line clocked at 10 kHz actually has a useful bandwidth of only 3.3 kHz.

Figure 6.3 shows the block diagram of a practical analogue delay-line system. The input signal is fed to the delay line via a low-pass filter with a cut-off frequency that is one third (or less) of the clock value, and is vital to overcome 'aliasing' or intermodulation problems. The delay line output is passed through a second low-pass filter, which also has a cut-off frequency one third (or less) of that of the clock and which serves the dual purposes of rejecting clock break-through signals and integrating the delay-line output

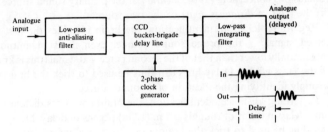

Figure 6.3 *Block diagram of a basic analogue delay-line system*

pulses, so that the final output signal is a faithful (but time-delayed) copy of the original.

We will take a closer look at various elements of the *Figure 6.3* circuit, and at some practical delay line chips, later in this chapter. In the meantime, let us digress slightly and look at an associated subject, which is known as psycho-acoustics.

Psycho-acoustics

Many of the special effects obtainable with delay lines rely on the human brain's idiosyncratic behaviour when interpreting sounds. Basically, the brain does not always perceive sounds as they truly are, but simply 'interprets' them so that they conform to a preconceived pattern; it can easily be fooled into misinterpreting the nature of sound. The study of this particular subject is known as psycho-acoustics. Here are some relevant psycho-acoustic 'laws' that are worth knowing:

1 If the ears receive two sounds that are identical in form but are time-displaced by less than 10 ms, the brain integrates them and perceives them as a *single* (undisplaced) sound.

2 If the ears receive two sounds that are identical in form but time-displaced by 10 to 50 ms, the brain perceives them as two *independent* sounds but integrates their information content into a single easily recognizable pattern, with no loss of information fidelity.

3 If the ears receive two signals that are identical in form but time-displaced by greater than 50 ms, the brain perceives them as two independent sounds but may be *unable* to integrate them into a recognizable pattern.

4 If the ears receive two sounds that are identical in basic form but not in magnitude, and which are time-displaced by more than 10 ms, the brain interprets them as two sound sources (primary and secondary) and draws conclusions concerning (a) the *location* of the primary sound source and (b) the relative *distances apart* of the two sources.

Regarding 'location' identification, the brain identifies the first perceived signal as the prime sound source, even if its magnitude is substantially lower than that of the second perceived signal (this is known as the Hass effect). Delay lines can thus be used to trick the brain into wrongly identifying the location of a sound source.

Regarding 'distance' identification, the brain correlates distance and time-delay in terms of roughly 0.3 m (12 in) per ms of delay. Delay lines can thus be used to trick the brain about 'distance' information.

5 The brain uses echo and reverberation (repeating echoes of diminishing amplitude) information to construct an image of environmental conditions, for example, if echo times are 50 ms but reverb time is 2 s, the brain may interpret the environment as being a 15 m (50 ft) cave or similar hard-faced structure, but if the reverb time is only 150 ms it may interpret the environment as being a wide softly furnished room. Delay lines can thus be used to trick the brain into drawing false conclusions concerning its environment, as in ambience synthesizers or room expanders.

6 The brain is very sensitive to brief transient increases in volume, such as caused by clicks and scratches on discs, but is blind to transient drops in volume. Delay lines can be used to take advantage of this effect in record players, where they can be used (on conjunction with other circuitry) to effectively predict the arrival of a noisy click/scratch and replace it with an 'unseen' neutral or negative transient.

Delay-line applications

Simple musical effects
Figures 6.4 to *6.16* show a variety of basic analogue delay-line applications. In these diagrams we have, for the sake of simplicity, ignored the presence of the usual input/output low-pass filters. Let us start by looking at some simple musical effects circuits.

Figure 6.4 shows how the delay line can be used to apply vibrato (frequency modulation) to any input signal. The low-frequency sinewave generator modulates the voltage controlled oscillator (VCO) clock generator frequency and thus causes the output signals to be similarly time-delay or 'frequency' modulated.

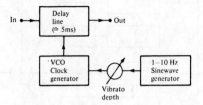

Figure 6.4   *True vibrato circuit, which applies slow frequency modulation to all input signals*

Figure 6.5 shows the delay line used to give a double-tracking effect. The delay time is in the 'perceptible' range 10 to 25 ms, and the delayed and direct signals are added in an audio mixer to give the composite 'two signals' output shown in the diagram. If a solo singer's voice is played through the unit, it sounds like a pair of singers in very close harmony. Alternative names for this type of circuit are mini-echo and micro-chorus.

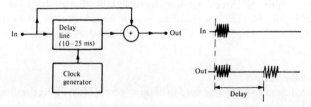

Figure 6.5   *Double-tracking circuit*

Figure 6.6 shows how the above circuit can be modified to act as an auto-double-tracking (ADT) or mini-chorus unit. Clock signals come from a VCO that is modulated by a slow oscillator, so that the delay times slowly vary. The effect is that, when a solo singer's voice is played through the unit, it sounds like a pair of singers in loose or natural harmony.

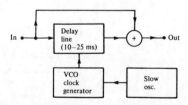

Figure 6.6   *Auto-double-tracking (ADT) or mini-chorus circuit*

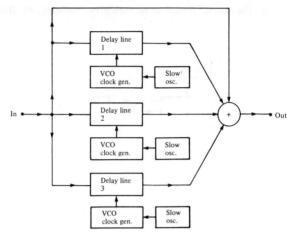

Figure 6.7   *'Chorus' generator*

Figure 6.7 shows how three ADT circuits can be wired together to make a chorus machine. All three delay-lines have slightly different delay times. The original input and the three delay signals are added together, the net effect being that a solo singer sounds like a quartet, or a duet sounds like an octet, etc.

Comb filter circuits
Figure 6.8 shows a delay line used to make a 'comb' filter. Here, the direct and delayed signals are added together; signal components that are in-phase when added give an increased output signal amplitude, and those that are in anti-phase tend to self-cancel and give a reduced output level. Consequently, the frequency response shows a series of notches, the notch spacing being the reciprocal of the line delay time (1 kHz spacing at 1 ms delay, 250 Hz spacing at 4 ms delay). These phase-induced notches are typically only 20 to 30 dB deep.

The two most popular musical applications of the comb filter are in phasers

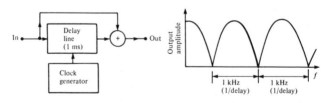

Figure 6.8   *CCD comb filter. Notches are about 20 to 30 dB deep, 1 kHz apart*

and flangers. In the phaser (*Figure 6.9*) the notches are simply swept slowly up and down the audio band via a slow-scan oscillator, introducing a pleasant acoustic effect on music signals.

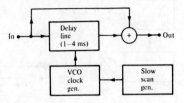

Figure 6.9 *A phaser is a variable comb filter, in which the notches are slowly swept up and down the audio band*

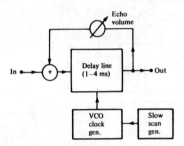

Figure 6.10 *A flanger is a phaser with accentuated and variable notch depth*

The *Figure 6.10* flanger circuit differs from the phaser in that the mixer is placed ahead of the delay line and part of the delayed signal is fed back to one input of the mixer, so that in-phase signals add together regeneratively. Amplitudes of the peaks depend on the degree of feedback, and can be made very steep. These phase-induced peaks introduce very powerful acoustic effects as they are swept up and down through music signals via the slow-scan oscillator.

Echo/reverb circuits
Figure 6.11 shows a basic echo unit, in which the direct and delayed signals are simply added together; the delay (echo) may vary from 10 ms to 250 ms and is usually adjustable, as also is the echo amplitude. Note that this circuit produces only a single echo.

The echo/reverb circuit of *Figure 6.12* produces multiple or repeating echoes (reverberation). It uses two mixers, one ahead of the delay line and the other at the output. Part of the delay output is fed back to the input mixer, so that the circuit gives echoes of echoes of echoes, etc. The reverb time is defined as the time taken for the repeating echo to fall by 60 dB relative to the original

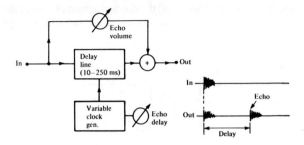

Figure 6.11 *An echo unit*

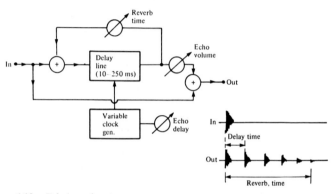

Figure 6.12 *Echo/reverb unit*

input signal, and depends on the delay time and the overall attenuation of the feedback signals. Echo delay time, echo volume and reverb time are all independently variable.

Figure 6.13 shows the basic circuit of an ambience synthesizer or room expander. Here, the outputs of a conventional stereo hi-fi system are summed to give a mono image and the resulting signal is then passed to a pair of semi-independent reverb units (which produce repeating echoes but not the original signal). The reverb outputs are then summed and passed to a mono PA system and speaker, which is usually placed behind the listener. The system effectively synthesizes the echo and reverb characteristics of a chamber of any desired size, so that listeners can be given the impression of sitting in a cathedral, concert hall, or small club house, etc., while in fact sitting in their own living room. Such units produce very impressive results.

There are lots of possible variations on the basic *Figure 6.13* circuit. In some cases the mono signal is derived by differencing (rather than summing) the stereo signals, thereby cancelling centre-stage signals and overcoming a

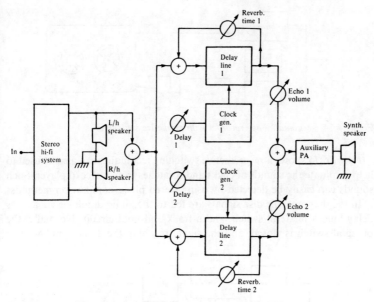

Figure 6.13 *Ambience synthesizer or room expander*

rather disconcerting 'announcer-in-a-cave' effect that occurs in 'summing' systems. The number of delay (reverb) stages may vary from one in the cheapest units to four in the more expensive.

Predictive switching circuits

Delay lines are particularly useful in helping to solve predictive or anticipatory switching problems, in which a switching action is required to occur slightly *before* some random event occurs. Suppose, for example, that you need to make recordings of random or intermittent sounds (thunder, speech, etc.). To have the recorder running continuously would be inefficient and expensive, and it would not be practical simply to try activating the recorder automatically via a sound switch, because part of the sound will already have occurred by the time the recorder turns on.

Figure 6.14 shows the solution to this problem. The sound input activates a sound switch, which (because of mechanical inertia) turns the recorder's motor on within 20 ms or so. In the meantime, the sound travels through the 50 ms delay line towards the recorder's audio input terminal, so that the recorder has already been turned on for 30 ms by the time the first part of the sound reaches it. When the original sound ceases, the sound switch turns off, but the switch extender maintains the motor drive for another 100 ms or so, enabling the entire 'delayed' signal to be recorded.

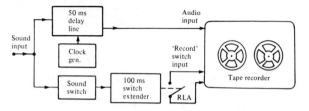

Figure 6.14   *Automatic tape recorder with predictive switching*

Figure 6.15 shows how predictive switching techniques can also be used to help eliminate the sounds of clicks and scratches from a record player. Such sounds can easily be detected by using stereo phase-comparison methods.

In the diagram, the disc signals are fed to the audio amplifier via a 3 ms delay line, a bilateral switch, and a track-and-hold circuit. Normally, the bilateral switch is closed, and the signal reaching the audio amplifier is a

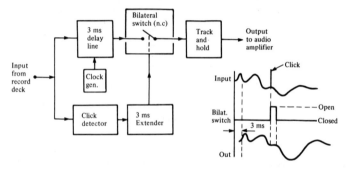

Figure 6.15   *Record (disc) click eliminator*

delayed but otherwise unmodified replica of the disc signal. When a click or scratch occurs on the disc, the detector/expander circuit opens the bilateral switch for a minimum of 3 ms, briefly blanking the audio feed to the amplifier. Because of the presence of the delay line, the blanking period effectively straddles the 'click' period, enabling its sound effects to be completely eliminated from the system (see psycho-acoustics).

Delay equalization
One of the most important applications of delay-line techniques is in sound-delay equalization of public addres systems in theatres and at open-air venues such as air shows, etc.

Sound travels through air at a rate of about 0.3 m (12 in) per ms. In simple multiple-speaker public address systems, in which all loudspeakers are fed

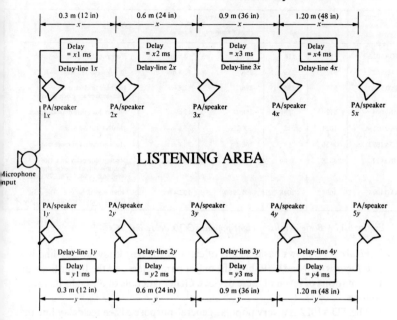

Figure 6.16 *Public address speech-delay equalization system*

with the same signal at the same moment, this factor inevitably creates multiple sound delays which can make voice signals unintelligible to the listener. This problem can be eliminated by delaying the signal feed to successive speakers (which are each driven by their own PA units) by progressive amounts (by 1 ms per 0.3 m of spacing from the original sound source), as shown in *Figure 6.16*. Ideally, the speakers should be spaced at 6 m (20 ft) intervals; the spacing interval must not be allowed to exceed 60 m (200 ft).

Practical ICs

Practical analogue delay-line ICs are available from several manufacturers, and usually have between 512 and 4096 bucket stages. Each IC can handle a wide range of clock signal frequencies and can thus provide a wide range of signal delays and bandwidths. They generate typically total signal distortion of about 1% and have typical signal-to-noise ratios of about 74 dB at maximum output. Most of them suffer some insertion loss or signal attenuation (usually a few dB) between the input and output of the IC.

Device number	Stages	Samples	Delay time (ms) v clock frequency	Delay at 7 kHz bandwidth	Notes
TDA 1022	512	256	256/f	12.8 ms	Very popular low-cost device
SAD 512	512	256	256/f	12.8 ms	Obsolescent 512-stage device
SAD 512D	512	256	256/2f	12.8 ms	Has built-in clock divider: uses single-phase clock input
MN 3004	512	256	256/f	12.8 ms	Modern high-performance device
SAD 1024A	1024	512	2 × 256/f	25.6 ms	Dual SAD512 device
TDA 1097	1536	768	768/f	38.4 ms	Modern general-purpose unit
MN 3011	3328	1664	1664/f	83.2 ms	Modern long-delay unit with multiple output taps (at stages 396, 662, 1194, 1726, 2790, and 3328).
SAD 4096	4096	2048	8 × 256/f	102.4 ms	4096-stage delay line: The clock terminal input capacitance = 1000 pF

Figure 6.17   *Basic details of eight popular CCD delay-line ICs*

Figure 6.17 shows basic details of eight of the best known delay-line ICs, and *Figures 6.18* and *6.19* show the major parameters and the IC outlines and pin notations of seven of these devices. General details of the eight ICs are as follows.

The TDA1022 is a very popular general-purpose 512-stage delay line that can give a wide range of delays (from 0.85 ms to 51.2 ms). It needs a 2-phase clock drive, and gives 12.8 ms at 7 kHz bandwidth when clocked at 20 kHz.

The SAD512 is an obsolescent 512-stage device that gives a performance similar to the TDA1022.

	TDA 1022	SAD 512D	MN 3004	SAD 1024A	TDA 1097	MN 3011	SAD4C
Supply voltage	−12 to −16 V	+10 to +17 V	−14 to −16 V	+10 to +17 V	−12 to −16 V	−14 to −16 V	+8 to +
Stages	512	512	512	2 × 512	1536	3328	4096
Clock frequency range (kHz)	5 to 3000	1 to 1000	10 to 100	1 to 1000	5 to 100	10 to 100	8 to 10
Signal delay range (ms)	0.85 to 51.2	0.26 to 200	2.56 to 25.6	0.26 to 100	7.7 to 153.6	16.6 to 166.4	2 to 25
Signal frequency range	DC−45 kHz	DC−200 kHz	DC−33 kHz	DC−200 kHz	DC−25 kHz	DC−20 kHz	DC−40
Maximum signal input volts	2 V rms	2 V p-p	1.8 V rms	2 V p-p	1.5 V rms	1 V rms	2 V p-p
Line attenuation	3.5 dB	2 dB	1.5 dB	0 dB	4 dB	0 dB	2 dB
Signal/noise ratio at max. output	74 dB	70 dB	85 dB	70 dB	77 dB	76 dB	70 dB
IC packaging	16-pin DIL	8-pin DIL	14-pin DIL	16-pin DIL	8-pin DIL	12-pin DIL	16-pin

Figure 6.18   *Major parameters of seven popular delay-line ICs*

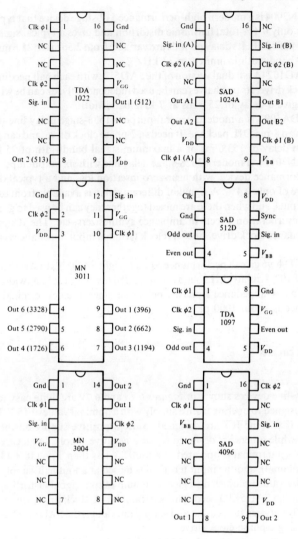

Figure 6.19   *Outlines and pin notations of seven popular delay-line ICs*

The SAD512D is an updated version of the SAD512. It has built-in output drivers and an integral clock divider that enables it to be driven by a single-phase clock input. This device offers an exceptionally wide range of delay times, and can give a signal bandwidth of up to 200 kHz.

The MN3004 is a modern high-performance 512-stage device that typically generates only 0.4% total harmonic distortion and gives a typical signal-to-noise ratio of 85 dB. Its delay periods are variable from 2.56 ms to 25.6 ms, and its signal bandwidth is limited to 33 kHz.

The SAD1024A is a dual version of the SAD512, with each half needing a 2-phase clock drive. Its two halves can be used independently, or can be wired in series to give a delay of 25.6 ms at 7 kHz bandwidth.

The TDA1097 is a modern general-purpose 1536-stage delay line that is housed in an 8-pin DIL package. It needs a 2-phase clock drive, and can give a maximum delay of 153.6 ms or a maximum signal bandwidth of 25 kHz.

The MN3011 is a modern 3328-stage delay line with six output taps. It is a high-performance device, with near-zero insertion loss and a typical distortion figure of only 0.4%. A signal of different delay is available from each of its six outputs, and when these are mixed together they can be used to generate natural reverberation effects in ambience synthesizers, etc. The IC needs a low-impedance clock drive, since its clock terminal input impedance is about 2000 pF.

The SAD4096 gives a performance equal to eight SAD512s in series; it gives a delay of 102.4 ms at 7 kHz bandwidth, or 250 ms at 3 kHz bandwidth. The IC needs a low-impedance 2-phase clock drive, since its clock terminal input capacitance is about 1000 pF.

Practical circuits

Delay-line circuits
The delay-line devices shown in *Figures 6.18* and *6.19* are quite easy to use. Each is designed to operate from a supply with a nominal value of 15 V. Note, however, that some ICs are designed to use a positive supply on the V_{DD} terminal, while others are designed to use a negative supply. In all cases, the V_{BB} (or V_{GG}) terminal is operated at about 1 V less than V_{DD} (at $+14$ V or -14 V nominal), and the input terminal is biased at about half-supply volts (the precise value is adjusted to give minimum output signal distortion). All ICs except the SAD512D need symmetrical 2-phase clock drive, which must switch fully between the ground and supply rail values; the SAD512D accepts a simple single-phase clock drive.

Figures 6.20 to *6.26* show how each of the above ICs can be wired as a simple delay-line circuit that uses a $+15$ V supply rail voltage and a grounded 'common' terminal. Note in the case of all ICs designed to operate with a negative V_{DD} voltage (the TDA1022, MN3004, TDA1097 and MN3011) that this is achieved by grounding the V_{DD} terminal and feeding the $+15$ V line to the Gnd pin.

In each of these circuits the input and output signals are applied and

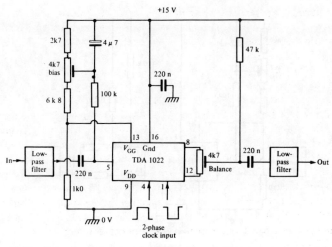

Figure 6.20   *Delay line using the TDA1022*

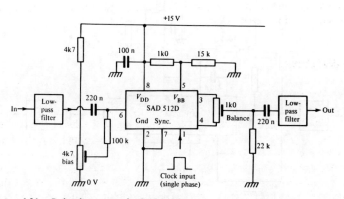

Figure 6.21   *Delay line using the SAD512D*

removed via low-pass filter stages; a pre-set is used to adjust the input DC bias so that symmetrical clipping occurs under overdrive conditions, and (except in the case of the MN3011) another pre-set is used to balance the IC's two outputs for minimum clock breakthrough.

Note in the *Figure 6.23* circuit that the two halves of the SAD1024A chip are wired in series, with the pin-5 output of delay-line A feeding to the pin-15 input of delay-line B; the unused pin-6 output of delay-line A is disabled by shorting it to pin-7.

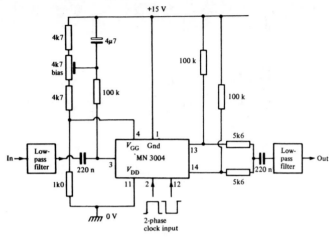

Figure 6.22   *Delay line using the MN3004*

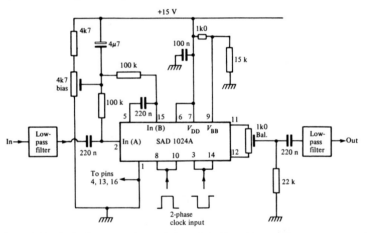

Figure 6.23   *Delay line using the two halves of the SAD1024A in the series-connected mode*

Clock generators

The clock signals to a CCD delay line should be reasonably symmetrical, with fast rise and fall times, and should switch fully between the supply rail voltages. *Figures 6.27* to *6.31* show a selection of suitable clock generators.

Clock generator design for the SAD512D delay line is very easy, since this IC incorporates a divider stage on its clock input line that enables the device to accept non-symmetrical single-phase clock signals.

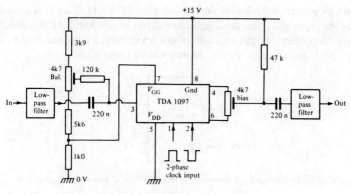

Figure 6.24   *Delay line using the TDA1097*

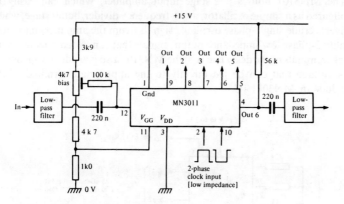

Figure 6.25   *Delay line using the MN3011*

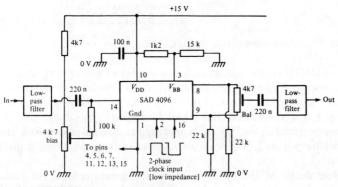

Figure 6.26   *Delay line using the SAD4096*

150   CCD audio delay-line circuits

Clock generator design for the MN3004 and MN3011 delay lines is made easy via a special 2-phase clock generator/driver IC, the MN3101. *Figure 6.27* shows the outline and pin notations of this device, which is housed in an 8-pin DIL package, and *Figure 6.28* shows its basic usage details.

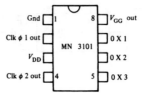

Figure 6.27   *Outline and pin notations of the MN3101 2-phase clock generator*

The MN3101 houses a 2-stage input amplifier, which can easily be configured as a simple oscillator, and a frequency divider/buffer stage, which converts crude single-phase oscillator signals (from the amplifier) into high-quality 2-phase low-impedance output signals that can be fed directly to the clock terminals of the delay line ICs; the MN3101 also provides a V_{GG} output bias voltage that can be fed directly to the appropriate terminal of the MN3004 or MN3011 IC.

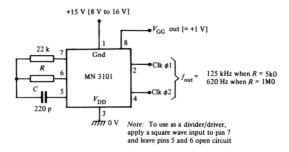

Figure 6.28   *Basic usage details of the MN3101*

The MN3101 can be used as a self-contained clock generator by connecting it as shown in *Figure 6.28*, in which case the output clock frequency is inversely proportional to the R value in the manner shown in the diagram. Alternatively, it can be used as a clock signal divider/buffer by removing the three components from pins 5 to 7 and connecting an external single-phase 'clock' signal to input pin 7; in this case the output clock frequency is half that of the input.

Note that the MN3101 can be used as a clock generator for any of the delay line ICs shown in this chapter. Alternatively, some popular CMOS digital ICs

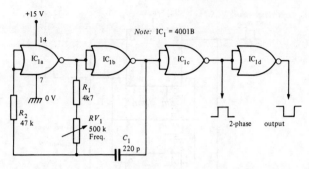

Figure 6.29   *Variable-frequency general-purpose 2-phase CMOS clock generator*

can be configured as excellent clock generators, and *Figures 6.29* to *6.31* show three practical circuits using such devices.

The general-purpose 2-phase generator of *Figure 6.29* uses an inexpensive 4001B IC; it can be used in most applications where a fixed or manually variable frequency is needed. The frequency can be swept over a 100:1 range via RV_1, and the centre frequency can be altered by changing the C_1 value.

The high-performance 2-phase generator of *Figure 6.30* is based on the voltage controlled oscillator (VCO) section of a 4046B chip, and is useful in applications where the frequency needs to be swept over a very wide range, or needs to be voltage controlled. The frequency is controlled by the voltage on pin 9, being at maximum (minimum delay) when pin 9 is high, and minimum (maximum delay) when pin 9 is low. Maximum frequency is determined by the C_2–R_1 values, and minimum frequency by the value of C_2 and the series values of R_2–RPS_1.

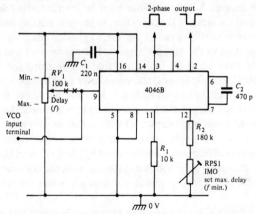

Figure 6.30   *High-performance voltage-controlled 2-phase CMOS clock generator*

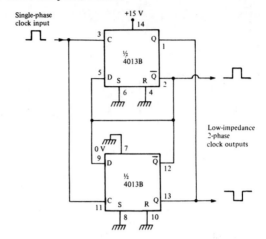

Figure 6.31   *Single-phase to 2-phase converter, with low impedance output*

The *Figures 6.29* and *6.30* circuits can be used to clock directly all CCD delay lines except the MN3011 and SAD4096, which have clock terminal capacitances of 1000 pF or more and need a low-impedance clock drive that is best provided by the *Figure 6.31* circuit, which uses both halves of a 4013B divider wired in parallel to give the required low-impedance 2-phase output; the circuit is driven by a single-phase clock signal, which can be obtained from either of the *Figures 6.29* or *6.30* circuits.

Filter circuits
In most applications, a low-pass filter must be inserted between the actual input signal and the input of the delay line, to prevent aliasing problems, and another in series with the output of the line, to provide clock-signal rejection and the integration of the 'sample' signals. For maximum bandwidth, both filters usually have a cut-off frequency that is one third (or less) of the maximum used clock frequency; the input filter usually has a 1st-order or better response, and the output filter has a 2nd-order or better response.

Figure 6.32 shows the practical circuit of a 25 kHz 2nd-order low-pass filter with ac-coupled input and output. The non-inverting terminal of the op-amp is biased at half-supply volts, usually by a simple potential divider network. The cut-off frequency can be varied by giving C_1 and C_2 alternative values, but in the same ratio as shown in the diagram; for example, cut-off can be reduced to 12.5 kHz by giving C_1 and C_2 values of 1n0 and 6n0 respectively.

Most delay lines suffer from a certain amount of insertion loss. Typically, if 100 mV is put in at the front of a delay line, only 70 mV or so appears at the

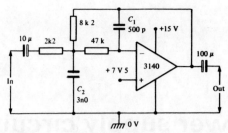

Figure 6.32   *A 25 kHz 2nd-order maximally-flat low-pass filter*

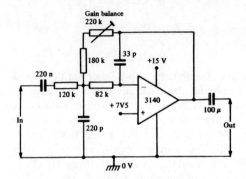

Figure 6.33   *Adjustable-gain 2nd-order low-pass output filter*

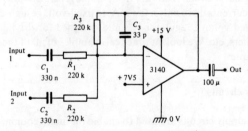

Figure 6.34   *Combined 2-input mixer/1st-order low-pass filter*

output. Often, the output low-pass filter is given a degree of compensatory gain, to give zero overall signal loss. *Figure 6.33* shows such a circuit, which has a nominal cut-off frequency of about 12 kHz, depending on the setting of the gain balance controls.

Finally, to complete this look at CCD delay line circuits, *Figure 6.34* shows how a 2-input unity-gain mixer (adder) can also be made to act as a 1st-order low-pass filter by simply wiring a roll-off capacitor (C_3) between the output and the inverting terminal of the op-amp. This type of circuit is often used at the front end of CCD flanger and reverberation designs.

7 Power supply circuits

Two frequent tasks facing the audio-IC user are those of designing basic power supplies to enable the audio equipment to operate from AC power lines, and of designing voltage regulators to power individual circuit sections at precise DC voltage values over wide ranges of load current variations.

Both of these design tasks are fairly simple. Basic power supply circuits consist of little more than a transformer-rectifier-filter combination, so the designer merely has to select the circuit values (using a few simple rules) to suit his/her own particular design requirements.

Voltage regulator circuits can vary from simple zener networks, designed to provide load currents up to only a few mA, to fixed-voltage high-current units using dedicated voltage regulator ICs and designed for driving hi-fi stereo power amplifiers, etc. We look at practical examples of all of these circuits in this final chapter.

Power supply circuits

Basic power supply circuits are used to enable the audio equipment to safely operate from the AC power lines (rather than from batteries), and consist of

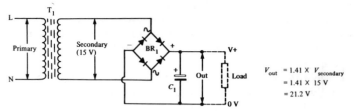

Figure 7.1   *Basic single-ended power supply using a single-ended transformer and bridge rectifier*

154

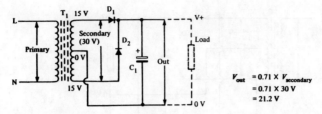

Figure 7.2 *Basic single-ended power supply using a centre-tapped transformer and two rectifiers*

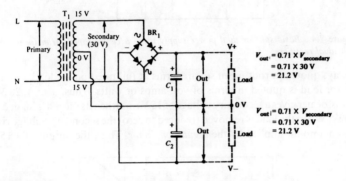

Figure 7.3 *Basic split or dual power supply using a centre-tapped transformer and bridge rectifier*

little more than a transformer that converts the ac line voltage into an electrically isolated and more useful ac value, and a rectifier-filter combination that converts this new ac voltage into smooth dc of the desired voltage value.

Figures 7.1 to *7.4* show the four most useful basic power supply circuits that the reader will ever need. The *Figure 7.1* circuit provides a single-ended dc supply from a single-ended transformer and bridge rectifier combination, and gives a performance that is virtually identical to that of the centre-tapped transformer circuit of *Figure 7.2*. The *Figures 7.3* and *7.4* circuits each provide split or dual dc supplies with nearly identical performances. The rules for designing these four circuits are very simple, as you will now see.

Transformer-rectifier selection

The three most important parameters of a transformer are its secondary voltage, its power rating, and its regulation factor. The secondary voltage is

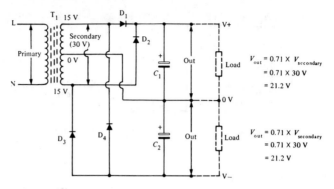

Figure 7.4 *Basic split or dual power supply using a centre-tapped transformer and individual rectifiers*

always quoted in root mean square terms at full rated power load, and the power load is quoted in terms of volt-amps or watts. Thus, a 15 V 20 VA transformer gives a secondary voltage of 15 V rms when its output is loaded by 20 W. When the load is removed (reduced to zero) the secondary voltage rises by an amount implied by the *regulation factor*. Thus, the output of a 15 V

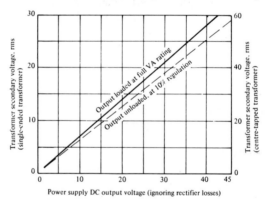

Power supply DC output voltage (ignoring rectifier losses)

Figure 7.5 *Transformer selection chart. To use, decide on the required loaded DC output voltage (say, 21 V), then read across to find the corresponding transformer secondary voltage (15 V single-ended or 30 V centre-tapped)*

transformer with a 10% regulation factor (a typical value) rises to 16.5 V when the output is unloaded.

Note here that the rms output voltage of the transformer secondary is *not* the same as the dc output voltage of the complete full-wave rectified power supply which, as shown in *Figure 7.5*, is in fact 1.41 times greater than that of a single-ended transformer, or 0.71 times that of a centre-tapped transformer (ignoring rectifier losses). Thus, a single-ended 15 V rms transformer with 10% regulation gives an output of about 21 V at full rated load (just under 1 A at 20 VA rating) and 23.1 V at zero load. When rectifier losses are taken into account the output voltages will be slightly lower than shown in the graph. In the two rectifier circuits of *Figures 7.2* and *7.4* the losses are about 600 mV, and in the bridge circuits of *Figures 7.1* and *7.3* they are about 1.2 V. For maximum safety, the rectifiers should have current ratings at least equal to the dc output currents.

Thus, the procedure for selecting a transformer for a particular task is quite simple. First, decide the dc output voltage and current that is needed; the product of these values gives the minimum VA rating of the transformer. Finally, consult the graph of *Figure 7.5* to find the transformer secondary rms voltage that corresponds to the required dc voltage.

The filter capacitor

The purpose of the filter capacitor is to convert the full-wave output of the rectifier into a smooth dc output voltage. The two most important parameters are its working voltage, which must be greater than the off-load output value of the power supply, and its capacitance value, which determines the amount of ripple that will appear on the dc output when current is drawn from the circuit.

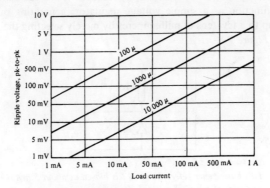

Figure 7.6   *Filter capacitor selection chart, relating capacitor size to ripple voltage and load current in a full-wave rectified 50–60 Hz powered circuit*

As a rule of thumb, in a full-wave rectified power supply operating from a 50 to 60 Hz power line, an output load current of 100 mA will cause a ripple waveform of about 700 mV pk-to-pk to be developed on a 1000 μF filter capacitor, the amount of ripple being directly proportional to the load current and inversely proportional to the capacitance value, as shown in the design guide of *Figure 7.6*. In most practical applications, the ripple should be kept below 1.5 V pk-to-pk under full load conditions. If very low ripple is needed, the basic power supply can be used to feed a 3-terminal voltage regulator IC, which can easily reduce the ripple by a factor of 60 dB or so at low cost.

Voltage regulator circuits

Practical voltage regulators may vary from simple zener diode circuits designed to provide load currents up to only a few milliamperes, to fixed- or variable-voltage high-current circuits designed around dedicated 3-terminal voltage regulator ICs. Circuits of all these types are described in the remainder of this chapter.

Zener-based circuits

Figure 7.7 shows how a zener diode can be used to generate a fixed reference voltage by passing a current of about 5 mA through it from the supply line via limiting resistor R. In practice, the output reference voltage is not greatly influenced by sensible variations in the diode current value, and these may be caused by variations in the values of R or the supply voltage, or by drawing current from the output of the circuit. Consequently, this basic circuit can be made to function as a simple voltage regulator, generating output load currents up to a few tens of milliamperes by merely selecting the R value as shown in *Figure 7.8*.

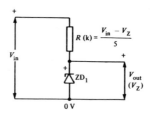

$$R\,(k) = \frac{V_{in} - V_Z}{5}$$

Figure 7.7   *This basic zener reference circuit is biased at about 5 mA*

Here, the value of R is selected so that it passes the maximum desired output current plus 5 mA. Consequently, when the specified maximum output load current is being drawn the zener passes only 5 mA, but when zero load current is being drawn it passes all of the R current, and the zener thus dissipates maximum power. Note that the power rating of the zener must not be exceeded under this 'no load' condition.

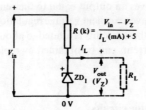

Figure 7.8 *This basic zener regulator circuit can supply load currents of a few tens of milliamperes*

The available output current of a zener regulator can easily be increased by wiring a current-boosting voltage follower circuit into its output, as shown in the series-pass voltage regulator circuits of *Figures 7.9* and *7.10*.

In *Figure 7.9*, emitter follower Q_1 acts as the voltage following current booster, and gives an output voltage that is 600 mV below the zener value under all load conditions; this circuit gives reasonably good regulation. In *Figure 7.10*, Q_1 and the CA3140 op-amp form a precision current-boosting

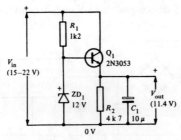

Figure 7.9 *This series-pass zener-based regulator circuit gives an output of 11.4 V and can supply load currents up to about 100 mA*

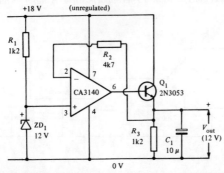

Figure 7.10 *This op-amp based regulator gives an output of 12 V at load currents up to 100 mA and gives excellent regulation*

voltage follower that gives an output equal to the zener value under all load conditions; this circuit gives excellent voltage regulation. Note that the output load current of each of these circuits is limited to about 100 mA by the power rating of Q_1; higher currents can be obtained by replacing Q_1 with a power Darlington transistor.

Fixed 3-terminal regulator circuits

Fixed-voltage regulator design has been greatly simplified in recent years by the introduction of 3-terminal regulator ICs such as the 78xxx series of positive regulators and the 79xxx series of negative regulators, which incorporate features such as built-in fold-back current limiting and thermal protection, etc. These ICs are available with a variety of current and output voltage ratings, as indicated by the 'xxx' suffix; current ratings are indicated by the first part of the suffix (L = 100 mA, blank = 1 A, S = 2 A), and the voltage ratings by the last two parts of the suffix (standard values are 5 V, 12 V, 15 V and 24 V). Thus, a 7805 device gives a 5 V positive output at a 1 A rating, and a 79L15 device gives a 15 V negative output at 100 mA rating.

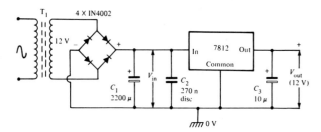

Figure 7.11   *Connections for using a 3-terminal positive regulator, in this case a 12 V 1 A 78 type*

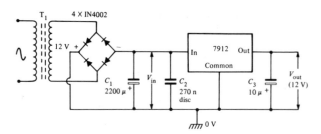

Figure 7.12   *Connections for using a 3-terminal negative regulator, in this case a 12 V 1 A 79 type*

Three-terminal regulators are very easy to use, as shown in the basic circuits of *Figures 7.11* to *7.13*, which show the connections for making positive, negative and dual regulator circuits respectively. The ICs shown are 12 V types with 1A ratings, but the basic circuits are valid for all other voltage values, provided that the unregulated input is at least 3 V greater than the

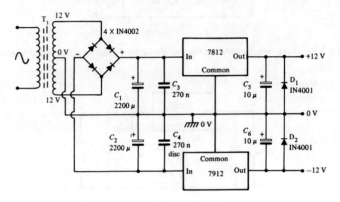

Figure 7.13   *Complete circuit of a 12 V 1 A dual power supply using 3-terminal regulator ICs*

desired output voltage. Note that a 270 nF or greater disc (ceramic) capacitor must be wired close to the input terminal of the IC, and a 10 μF or greater electrolytic is connected across the output. The regulator ICs typically give about 60 dB of ripple rejection, so 1 V of input ripple appears as a mere 1 mV of ripple on the regulated output.

Voltage variation

The output voltage of a 3-terminal regulator IC is actually referenced to the IC's common terminal, which is normally (but not necessarily) grounded. Most regulator ICs draw quiescent currents of only a few mA, which flow to ground via this common terminal, and the IC's regulated output voltage can thus easily be raised above the designed value by simply biasing the common terminal with a suitable voltage, making it easy to obtain odd-ball output voltage values from these fixed voltage regulators. *Figures 7.14* to *7.16* show three ways of achieving this.

In *Figure 7.14* the bias voltage is obtained by passing the IC's quiescent current (typically about 8 mA) to ground via RV_1. This design is adequate for many applications, although the output voltage shifts slightly with changes in

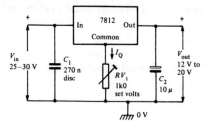

Figure 7.14   *Very simple method of varying the output voltage of a 3-terminal regulator*

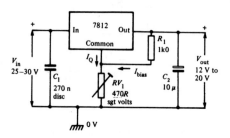

Figure 7.15   *An improved method of varying the output of a 3-terminal regulator*

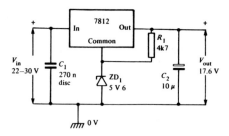

Figure 7.16   *The output voltage of a 3-terminal regulator can be increased by a fixed amount by wiring a suitable zener diode in series with the common terminal*

quiescent current. The effects of such changes can be minimized by using the circuit of *Figure 7.15*, in which the RV_1 bias voltage is determined by the sum of the quiescent current and the bias current set by R_1 (12 mA in this example). If a fixed output with a value other than the designed voltage is required, it can be obtained by wiring a zener diode in series with the common terminal as shown in *Figure 7.16*, the output voltage then being equal to the sum of the zener and regulator voltages.

Current boosting

The output current capability of a 3-terminal regulator can be increased by using the circuit of *Figure 7.17*, in which current boosting can be obtained via bypass transistor Q_1. Note that R_1 is wired in series with the regulator IC. At low currents insufficient voltage is developed across R_1 to turn Q_1 on, so all the load current is provided by the IC. At currents of 600 mA or greater sufficient voltage (600 mV) is developed across R_1 to turn Q_1 on, so Q_1 provides all current in excess of 600 mA.

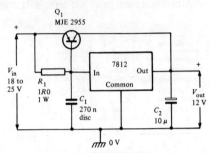

Figure 7.17 *The output current capacity of a 3-terminal regulator can be boosted via an external transistor. This circuit can supply 5 A at a regulated 12 V*

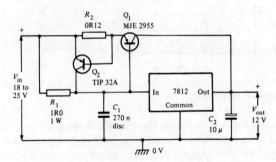

Figure 7.18 *This version of the 5 A regulator has overloaded protection provided via Q_2*

Figure 7.18 shows how the above circuit can be modified to provide the bypass transistor with overload current limiting via 0.12 Ω current-sensing resistor R_2 and turn-off transistor Q_2, which automatically limit the output current to about 5 A.

Variable 3-terminal regulator circuits

The 78xxx and 79xxx range of 3-terminal regulator ICs are designed for use in fixed-value output voltage applications, although their outputs can in fact be varied over limited ranges. If readers need regulated output voltages that are variable over very wide ranges, they can obtain them by using the 317 K or 338 K 3-terminal variable regulator ICs.

Figure 7.19 shows the outline, basic data and the basic variable regulator circuit that is applicable to these two devices, which each have built-in foldback current limiting and thermal protection and are housed in TO3 steel packages. The major difference between the devices is that the 317 K has a 1.5 A current rating compared to the 5 A rating of the 338 K. Major features of both devices are that their output terminals are always 1.25 V above their adjust terminals, and their quiescent or adjust-terminal currents are a mere 50 μA or so.

Parameter	317 K	338 K
Input voltage range	4–40 V	4–40 V
Output voltage range	1.25–37 V	1.25–32 V
Output current rating	1.5 A	5 A
Line regulation	0.02 %	0.02 %
Load regulation	0.1 %	0.1 %
Ripple rejection	65 dB	60 dB

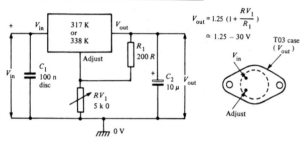

Figure 7.19   *Outline, basic data and application circuit of the 317 K and 338 K variable-voltage 3-terminal regulators*

Thus, in the *Figure* 7.19 circuit, the 1.25 V difference between the 'adjust' and output terminals makes several milliamperes flow to ground via RV_1, thus causing a variable adjust voltage to be developed across RV_1 and applied to the adjust terminal. In practice, the output of this circuit can be varied from 1.25 to 33 V via RV_1, provided that the unregulated input voltage is at least 3 V greater than the output. Alternative voltage ranges can be obtained by using other values of R_1 and/or RV_1, but for best stability the R_1 current should be at least 3.5 mA.

The basic *Figure 7.19* circuit can be usefully modified in a number of ways; its ripple rejection factor, for example, is about 65 dB, but this can be increased to 80 dB by wiring a 10 μF bypass capacitor across RV_1, as shown in *Figure 7.20*, together with a protection diode that stops the capacitor discharging into the IC if its output is short-circuited.

Figure 7.21 shows a further modification of the *Figure 7.20* circuit; here, the transient output impedance of the regulator is reduced by increasing the C_2 value to 100 μF and using diode D_2 to protect the IC against damage from the stored energy of this capacitor if an input short occurs.

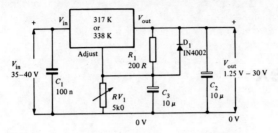

Figure 7.20   *This version of the variable-voltage regulator has 80 dB of ripple rejection*

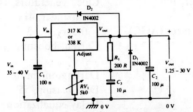

Figure 7.21   *This version of the regulator has 80 dB ripple rejection, a low impedance transient response, and full input and output short-circuit protection*

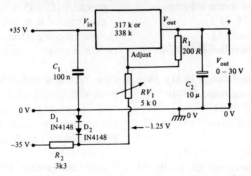

Figure 7.22   *The output of this version of the regulator is fully variable from 0 to 30 V*

Finally, *Figure 7.22* shows how the circuit can be modified so that its output is variable all the way down to zero volts, rather than to the 1.25 V of the earlier designs. This is achieved by using a 35 V negative rail and a pair of series-connected diodes that clamp the low end of RV_1 to minus 1.25 V.

Index

Integrated circuits (IC) by type number